BUS

Inside the Sm

Springer
London
Berlin
Heidelberg
New York
Hong Kong
Milan
Paris
Tokyo

Richard Harper (Ed.)

Inside the Smart Home

Springer

Richard Harper, PhD
The Appliance Studio, University Gate East, Park Row, Bristol BS1 5UB, UK

British Library Cataloguing in Publication Data
Inside the smart home
  1. Home automation 2. Home automation–Social aspects
  3. Household appliances–Automatic control 4. Household electronics
  I. Harper, Richard, 1960–
643.6
ISBN 1852336889

Library of Congress Cataloging-in-Publication Data
Inside the smart home / Richard Harper (ed.).
   p. cm.
  ISBN 1-85233-688-9 (alk. paper)
  1. Home automation. 2. Households. 3. Lifestyles.
  4. Domestic engineering. I. Harper, Richard, 1960–
TK7881.25.I575 2003
643′.6–dc21   2003050547

ISBN 1-85233-688-9 Springer-Verlag London Berlin Heidelberg
a member of BertelsmannSpringer Science+Business Media GmbH
www.springer.co.uk

Typeset by Florence Production Ltd, Stoodleigh, Devon
Printed and bound in the United States of America
34/3830-543210 Printed on acid-free paper SPIN 10890627

For Neville Moray

Contents

List of Contributors xi

1 **Inside the Smart Home: Ideas, Possibilities and Methods** 1
   Richard Harper
   1.1 Introduction 1
   1.2 An Introduction to the History of User Research in this
       Area 5
   1.3 Structure of the Book 7

Part 1   Conceptions of the Home 15

2 **Smart Homes: Past, Present and Future** 17
   Francis K. Aldrich
   2.1 Introduction 17
   2.2 Past History of the Smart Home 18
   2.3 Present Status of the Smart Home 22
   2.4 Future Prospects for the Smart Home 27
   2.5 Conclusion 36

3 **Households as Morally Ordered Communities:**
   Explorations in the Dynamics of Domestic Life 41
   John D. Strain
   3.1 Preamble 41
   3.2 Theoretical Context 41
   3.3 Analysing the Practices – Continuity and Change 47
   3.4 Changing the Paradigm of Banking 52
   3.5 Conclusion 59

4 **Time as a Rare Commodity in Home Life** 63
   Lynne Hamill
   4.1 Introduction 63
   4.2 The Economics of the Household 63
   4.3 Technology in the Home 66
   4.4 Case Study: TVs, VCRs and DVDs 70
   4.5 Analysis 74

5   **Emotional Context and "Significancies" of Media** 79
     Sue Peters
     5.1   Introduction 79
     5.2   Television 80
     5.3   PC/Internet 85
     5.4   Mobile as Hybrid 87
     5.5   Media Streaming 89
     5.6   Routine Activity 90
     5.7   Regions and Blurring 92
     5.8   Conclusions 94

Part 2   Designing for the Home 99

6   **Paper-mail in the Home of the 21st Century** 101
     Richard Harper and Brian Shatwell
     6.1   Background 101
     6.2   An Overview of Findings 104
     6.3   Conclusions 111

7   **Switching On to Switch Off** 115
     Alex Taylor and Richard Harper
     7.1   Introduction 115
     7.2   Findings 116
     7.3   Analyses of the Three Types of Viewing 119
     7.4   Lessons for Design 122
     7.5   Conclusions 124

8   **The Social Context of Home Computing** 127
     David Frohlich and Robert Kraut
     8.1   Introduction 127
     8.2   The Use of Domestic Time 128
     8.3   The Use of Domestic Space 131
     8.4   The Use of Domestic Technology 134
     8.5   Methods 136
     8.6   Results 137
     8.7   Power, Regulation and Control 151
     8.8   Discussion 155

9   **Design with Care: Technology, Disability and the Home** 163
     Keith Cheverst, Karen Clarke, Guy Dewsbury, Terry Hemmings,
     John Hughes and Mark Rouncefield
     9.1   Introduction: Sociology, the Home, Design and Disability ... 163
     9.2   Eliciting Design Requirements for Domestic Environments .. 168
     9.3   Research Methods for Design for Domestic Environments ... 169
     9.4   Design With Care: Moving Towards Appropriate Design 175

Part 3   The Home of the Future 181

**10   Towards the Unremarkable Computer: Making Technology
      at Home in Domestic Routines** 183
      *Peter Tolmie, James Pycock, Tim Diggins, Allan MacLean
      and Alain Karsenty*
      10.1 Introduction 183
      10.2 Work Routines 186
      10.3 Returning to the Home 187
      10.4 Routine Instances 188
      10.5 The Alarm Clock 190
      10.6 Ubiquitous Computing and the Quest for the "Invisible" 197
      10.7 Conclusion 204

**11   Daily Routines and Means of Communication in a
      Smart Home** 207
      Sanna Leppänen and Marika Jokinen
      11.1 Introduction 207
      11.2 Daily Routines Structure the Everyday Life 208
      11.3 Family Communications 211
      11.4 Traditional Paper Versus Electronic Communication 215
      11.5 Emotions and Smart Home Technology 219
      11.6 Conclusions 221

12   Living Inside a Smart Home: A Case Study 227
      Dave Randall
      12.1 CSCW and the Domestic Environment 227
      12.2 The Study 230
      12.3 Social Connectivity 234
      12.4 Conclusion 241

**13   Smart Home, Dumb Suppliers? The Future of Smart
      Homes Markets** 247
      James Barlow and Tim Venables
      13.1 Introduction 247
      13.2 "Smart Homes": A Recent History 248
      13.3 New Players and New Markets? 250
      13.4 Networks and Transmission Technologies 253
      13.5 Models for the Future 256
      13.6 Future Challenges 257
      13.7 Conclusions 260

Index 263

Contributors

Frances K. Aldrich, Department of Psychology, School of Life Sciences, University of Sussex, UK

James Barlow, Innovation Studies Centre, Imperial College London, UK

Keith Cheverst, Department of Computing, Lancaster University, UK

Karen Clarke, Department of Computing, Lancaster University, UK

Guy Dewsbury, Department of Computing, Lancaster University, UK

Tim Diggins, Image Semantics Ltd., Cambridge, UK

David Frohlich, Hewlett Packard Labs, Bristol, UK

Lynne Hamill, Digital World Research Centre, University of Surrey, UK

Richard Harper, Appliance Studio, Bristol, UK

Terry Hemmings, School of Computer Science and Information Technology, Nottingham University, UK

John Hughes, Department of Sociology, Lancaster University, UK

Marika Jokinen, Digital Media Institute, Tampere University of Technology, Finland

Alain Karsenty, *formerly of* Xerox Research Centre, Europe, Grenoble, France

Robert Kraut, Carnegie Mellon University, Pittsburgh, USA

Sanna Leppänen, Digital Media Institute, Tampere University of Technology, Finland

Allan MacLean, Image Semantics Ltd., Cambridge, UK

Sue Peters, Teleconomy, Lancaster, UK

James Pycock, Image Semantics Ltd., Cambridge, UK

Dave Randall, Department of Sociology, Manchester Metropolitan University and Department of Human Work Science, Blekinge Institute of Technology, UK

Mark Rouncefield, Department of Computing, Lancaster University, UK

Brian Shatwell, Royal Mail, UK

John D. Strain, Digital World Research Centre, University of Surrey, UK

Alex Taylor, Digital World Research Centre, University of Surrey, UK

Peter Tolmie, Xerox Research Centre, Europe, Grenoble, France

Tim Venables, Innovation Studies Centre, Imperial College London, UK

Inside the Smart Home: Ideas, Possibilities and Methods

Richard Harper

1.1 Introduction

In 2001, Orange, a UK mobile network operator, announced the "Orange at Home" project, a smart house incorporating the latest technology wizardry built some 20 miles north of London. It was intended to be more than a mere showcase, with plans for real families to move in and live with the smart home. My then research establishment, the Digital World Research Centre at the University of Surrey, was commissioned to study how these families reacted to their new home, and to report lessons for the future development of smart homes and smart home technologies.

Why would a mobile network operator want to build a smart house? And what is a smart house anyway? In this section I want to introduce the reader to smart homes by way of a quick history, outline some of the research approaches to smart homes before saying something about why a mobile operator would want to build one. Having then said something about the kinds of technologies that are being piloted in smart home environments, I will outline the structure of the book that is presented here, starting as it does with a detailed history of the concept of smart homes and ending with an analysis of those organisations that supply all that goes in to make a smart home.

1.1.1 A History

Although many of us will be vaguely familiar with the term "smart house", few of us will have a very concrete understanding of what it means. It was first used in an official way as long ago as 1984 by the American Association of House Builders, though the first "wired homes" were actually built by hobbyists in the early 1960s. And this development is key to what is meant by smart homes. For a home is not smart because of how well it is built, nor how effectively it uses space; nor because it is environmentally friendly, using solar power and recycling waste water, for example. A smart home may, and indeed often does, include these things, but what makes it smart is the interactive technologies that it contains.

Now, leaving aside the fact that in the early 1960s there was not much interactive technology about – after all the famous experiments on the mouse and the pointer at the Stanford Research Institute (SRI) weren't until 1967 – since that time and right up to the present there has been very little take-up of smart homes. They have not been a hit because they have been too expensive, the housing stock is old, there has been a tendency for little networked connectivity, and finally, there has been too much technology push, and little attention given to users or usability.

Now focusing on the last of these, usability, what one finds is that the design of "domestic technology" has been something of a Cinderella science. The principal reasons for this neglect are:

1. a lack of motivation to increase productivity in domestic work;
2. little involvement of users of the technology in the design process;
3. the view held by product designers that domestic technology is unexciting;
4. a continued focus on stand-alone appliances in the design of new technology.

This situation does seem to be changing, though there are peculiar problems when trying to design interactive technologies for the home. After all, it is not like a workplace where you get planning, maintenance and – most important of all – technical support. Families, after all, are not structured like organisations. To make matters worse, "users" go from babies to old age pensioners. And finally, home users are hard to study anyway: what family would want observers hanging around all day and (perhaps) all of the night?

This is not to say it is impossible to design effectively for the home. But it is to say that it is hard. Notwithstanding these difficulties, there is a great deal of interest in smart homes at the present time. This is attested to by the Orange at Home project, as well as the increasing amount of commercial and academic research around the world.

1.1.2 Smart Homes Today

What then, does a smart house at the start of the 21st century have in it? What makes it smart? Let's take the Orange at Home example. Here, a 50- year-old house has been wired up with a network, run by a server that operates all of the functions of the house. Lighting, heating, security, audio-visual systems, curtains, baths and numerous other appliances can be controlled through WAP, SMS or a dial tone on a mobile phone (an Orange one obviously!); and "wirefree" technology allows PDAs and webtablets within the house to do the same. There are also ordinary PCs, though what makes these a little out of the ordinary is their connection to broadband networks. Finally, there are various other technologies like a health monitoring system in the house's "sport room".

The purpose of Orange's investment is to provide an opportunity to explore what users may want by giving them as many opportunities as possible with current off-the-shelf technology, even though it was recognised from the outset that some of the technologies would not be well received.

What was learnt when families were put in the Orange at Home environment was that technologies that succeed in work environments sometimes fail in home settings. This was not because they offer the wrong solutions but because what is important is different. The house provided wall panel units for controlling lighting and other functions, for example, and these are fairly standard in current office settings. But in the home, users found the functions overly complex if not unnecessary, and were much happier with simpler control devices like the old-fashioned rocker switch. Another finding, again not so surprising, was that issues of conflict showed themselves over who controlled the systems: kids would regularly override their Mum's selection of music on the centrally controlled audio-system, for example.

In many ways these were expected problems and concerns, and have since been dealt with. More interestingly, there were issues to do with where access to such things as online shopping services was provided within the house: the kitchen would seem obvious, but the PCs that could provide access to those services were not designed for such an environment, and were located in the smart home's office. They were still used for online shopping, but that required the householder to go to the 'wrong place' for the task. In other words, PC technology may provide all that users need for an office desktop, but in the home certain aspects of the same technology inhibit what might be an ideal solution for users' needs.

Designing PCs for kitchens may seem prosaic, but it is one of the great hopes of smart homes that the technology they provide will achieve a blurring of previously existing barriers and thresholds. In this example, we are thinking of being able to order food when and where the need for it is discovered, and that will most probably be in the kitchen. Other thresholds also come to mind of course, the most obvious being between home and work but where "home work" is done in the house, and conversely home specific activities like grocery management, needs to be though about carefully. The location of the interactive devices needs to be related to the patterns of space usage in the home.

Yet this is not the only important issue in making the technology fit in to home settings. There is also the question of interaction mode. Using hand held devices allowed family members mobility within the smart house, but often click throughs were perceived as overly complex: users preferred to use the volume knob on the hi-fi rather than the stylus input mode on their PDAs, for example. There is also the issue of using novel modes for input: here "hands free" commands come to mind, and the possibility of using voice activated controls for certain tasks where the hands are "tied up". Consider the predicament of wanting to adjust the heat on a hob when one is cutting garlic, for instance: here voice could do (as it were) the talking to the systems. And indeed many of those who stayed at the smart home expressed keen interest in such applications. Orange have been exploring this, but although interactive voice activated applications are becoming increasingly common, they have a long way to go before they can process the myriad commands that any cook would want. So it will be some time before smart homes provide the kind of solution that the film *2001: A Space Odyssey* brings to mind.

And this leads us on to the distance between users' hopes and reality, even the reality of 20 or 30 years hence. For what one can do with smart technologies is process information, but one cannot undertake all of the mechanics of home life. Dishwashers actually do the washing, but the machines still need loading and unloading; likewise the washing machine. This will not change since it is very unlikely that automated processes will emerge that will provide solutions that are either practical or cost effective for these kinds of tasks. People might like the idea of home robots doing the laundry, but it is not realistic.

In any case, what our studies did show is that what people want interactive technologies to provide is not automation, so much as communication, or as we like to put it, social connectivity. And this is the main reason why a mobile network operator like Orange is investing in smart homes. With the Internet, for example, families spread around the globe can and do set up web conferences, and though the bandwidths currently available may produce pretty grainy and poor quality images, a great deal of value is thereby provided. With new screen and tablet technologies, combined with air-based networks, the ways in which such needs can be supported offers many interesting new possibilities. This will be good news for those who provide the networks for such needs, such as Orange, as well as for the so-called terminal manufacturers who will produce the hardware.

1.1.3 The Future

Many designers are now trying to provide social connectivity solutions, and hopefully the Orange at Home project will incorporate some of them as they appear. But there are some fundamentals that need to be resolved if smart homes of the future are to succeed.

Perhaps most obviously there is a need for industry-wide standards that will allow the exchange of information and commands between various interactive technologies. Currently, most technologies communicate via proprietary protocols and this inhibits seamless interaction between technologies from different manufacturers. Though there are various on-going attempts to create standards both for smart homes and for hand-held devices more generally, the likelihood that agreements in this area will be reached soon is doubtful. It is not only there that large commercial interest is at play, but there are also complex technical and usability issues that have yet to be solved. The failure of Hewlett-Packard's JetSend technology is an instance of a standard that could not get past competitive commercial interests; the current technical and usability difficulties with Bluetooth an instance of the latter. It may be that academics will find a role here, both in terms of brokering standards, and also with technology through inventing something similar to HTML – though this time it won't be to share and read documents within physics laboratories, but for the home. (Just to remind the reader, HTML is the basic tagging language that has allowed the emergence of the World Wide Web. The language was invented by an academic at an EU research lab).

Related to these issues is the emergence of homes that are wired for sight and sound in much the same way that current homes are wired for electricity. It is worth recalling that when electricity firms first provided access to power, they assumed that householders would only want one point of access, not many plugs and sockets and power points. Once this had been realised, consumers of electricity were then provided with opportunities to appropriate the power source as they saw fit. Likewise there is a view that smart homes need to provide similar networks, though whether they combine fixed wiring or air-based facilities is neither here nor there. But given what we have said, it would seem unlikely that there will appear a single network solution to this need, and much more probable that the householder will be confronted with hybrid and mixed networks in the future. Consequently, these networks and the technologies they support will not necessarily be able to communicate with all the other technologies in the home, and will probably develop in such a fashion that closely related ones are linked through proprietary controls. There will also be a mix of so-called point solutions and generic ones, all in one way or another reliant on the various networks.

1.2 An Introduction to the History of User Research in this Area

Without wanting to say anything more about the Orange at Home project in this introduction – after all a later chapter deals with it in greater depth – what should have been conveyed thus far is the fact that designing for the smart home has its difficulties. One reason for this has to do with how little effort has been put into understanding what is needed. From any review of the smart home domain, it will be apparent that the design and use of technology within a domestic setting has been neglected by academia, and to a large extent by industry too. As Hindus (1999) points out, there is not yet a "critical mass" of interest in this area. In the mean-time, those looking for guidance from the research literature must pick their way through a fragmented area, gleaning what they can, where they can – from fields as diverse as sociology, ethnography, feminist analysis, human-computer interaction (HCI), computer-supported co-operative work (CSCW), artificial intelligence, buildings research, and health care. Issues for investigation, methodologies, research paradigms, and frame-works for analysis must all be decided upon with little guidance from the literature; there is even a lack of a "body of evidence" on the design and use of domestic technology which would enable some grounded exploration of these issues.

The apparent void in this area does not mean that researchers have to start from a blank sheet, however. As I have noted in a recent editorial of *Personal Technologies* (Harper 2000), it is clear that the technological models traditionally used to determine the design and likely role of tech-nologies can be and will need to supplemented by human sciences investigations into the behavioural patterns and needs of domestic users. The need to supplement technological visions of the home in the future is now well known (first stated as long ago as Kling, 1980), but the diffi-culties entailed in doing so are far from trivial.

Take sociological research into domestic life. It is both broad and extensive. Unfortunately, much, though not all, of this research is too theoretical to guide the design of technology or to understand the social shaping and impact of interactive technologies at point of use – in this case, in the domestic setting. For example, Nippert-Eng's *Home and Work* (1997) is one of the most recent monographs into the sociology of tech-nology in the home. Its main concern is not with the technology itself or its use, however, so much as what the role of that technology tells sociological theoreticians about the segregation of society into different "cognitive domains". The same concern with theoretical arguments holds true in the work of Castells (1985, 1998) and many others. In effect, tech-nology disappears from view.

This is not to say that undertaking such theoretically driven work in sociology is without its merits, but it is to say that it is limited in the insight it provides into the design or service functionality of current or future technology. In other words, such research may not be ideal for effective design. Such doubts arise because recent developments in the parallel (though obviously distinct) area of CSCW design has shown that such research (i.e. seeking theoretical abstraction), though interesting in its own right, does not generate the kind of rich, detailed descriptions that make design reasoning possible. Without doing so, it is difficult to see how sociological research can genuinely impact interactive system design or how it can be used to help predict the social shaping that such technology will go through.

In much of the literature in CSCW and to a lesser extent, OIS (office information system) and HCI, the view is expressed that rich descriptions call for sociological ethnography and particularly ethnomethodologically informed research (Harper et al., 2000; Hughes et al., 1998). As yet, no similar area of research has been developed in relation to the domestic environment (see Venkatesh, 1986, 1996b). But it may be premature to opt for one particular paradigm, albeit one that has shown some success. Given the dirth of research in the area, it is perhaps more appropriate to explore, both conceptually and theoretically, different approaches to the design, use and shaping of technologies in home environments. This might be combined with a broad view as to what "domestic technologies" consist of. After all, with the increasing use of embedded computing in all sorts of "white goods" in the home, as well as a spread of interest in smart home technologies which combine a variety of interactive applications with fairly prosaic services like water and heating provision, just what is meant by design is broad indeed.

1.3 Structure of the Book

It is in this context that this collection has been brought together. It consists of three parts, the first of which is *conceptions of the home*. The purpose of this section will be to address the issue of how to conceive of the home and hence the factors of relevance for design. The opening chapter in this section will provide a historical sketch of the dramatic revolution in domestic technology that has occurred over the past century. Whereas most domestic appliances available at the turn of the twentieth century would easily have been recognised by previous generations, the next hundred years has changed the scene unrecognisably. First, the introduction of electricity into homes in the first quarter of the century has provided a new source of clean, convenient power for appliances and spurred the introduction of hitherto unheard of equipment for the home. Secondly, the introduction of information technology into homes in the last quarter of the century opened up possibilities for exchanging information between people, appliances, systems and networks which we have still to fully explore. This chapter will overview these developments and attempt to characterise what future interactive smart home technologies might look like. Using lessons from the past and from other research reported in this book, a schema will be presented that defines the types of technology that are available for home consumption as well as those technologies that will create what have come to be called smart homes. A further schema presenting the present and future of smart house technology will also be presented.

The next chapter will then explain that homes are not merely repositories of human action nor can they be understood as the output of larger sociological phenomena, as is often argued in the mainstream sociological literature (see, for example, Haddon, 1992); Rather, households (or families or groups of people sharing common facilities –

however you want to label them) are in many ways communities in their own right: they possess a tradition, a moral order which frames and guides behaviour as well as the use of household facilities and technologies. This chapter will explore, using ethnographic evidence drawn from a panel of households in the south-east of England, how households exhibit these patterns of distinctive moral ordering. It will explore the problems of understanding them analytically; and on this basis, point towards how this moral ordering needs to factored in when trying to design and implement digital technologies in and for home settings. A particular theme will be to explore, on the basis of evidence, how the adoption of digital technologies will, as a matter of course, generate various conflicts, but not necessarily of power and hierarchy, having to do much more often with such everyday matters as "appropriate behaviours" for different times and places. Here issues to do with how the home is at once a place for hosting friends and for resting, for living out personal desires and for sharing with others who may have different desires, need to be managed and "worked out". Though these matters may seem from certain views prosaic, the conflicts that can result are often consequential. More importantly, they also have implications for what are the factors affecting the suitability of various digital technologies.

Morality is only one aspect of the home of course; albeit very important. Another is time. In the next chapter time will be considered as a rare commodity in home life and the implications of this explored. It should be clear that economic considerations should be fundamental to the design process, though it is more or less excluded from consideration not only for home technologies but for most interactive system design processes. The chapter will show how modern economics provides a useful tool to analyse consumer demand for new domestic technology, though doing so should not be confined to the usual metric, namely financial cost. This chapter sets out how an economics of the opportunity costs of time can be used to explain the success or failure of various types of new technologies in the home. Examples of technologies discussed include household entertainment technologies such as hi-fi and television, as well as the Internet for purchasing groceries and durables. Data for this are taken from UK and European sources from the 1950s to the present day.

As more channels to market reach higher levels of penetration, content providers are struggling to find the killer content, and are likewise struggling to understand the relationship between content and different channels. At times, it appears that basic rules of marketing have been forgotten and content providers seem unsure about what message works best on what medium. The next chapter provides an alternative view of consumer behaviour by concentrating on how the relationships and interactions with media create particular and local (or house specific) meaning systems. Drawing on a range of resources including Roland Barthes' writings on semiology and Veblen's theory of the leisure class, this chapter will examine why certain meanings and associations are attributed to various media in the home environment. Among other factors, historical patterns of use and assumptions about the social implications of media use (for example, assumptions around the TV as a box in the corner) have implied that no active relationship is associated with the device. This chapter will show that the passivity of the medium may indeed be a barrier to using interactive TV, and other interactive technologies that are beginning to invade the home.

The next section of the book, *designing for the home,* is focused more expressly on design. The first chapter in this section explores how the home is held together through communicative practices, and focuses in particular on the use of paper mail as a tool for domestic management. Reporting an ethnographically informed study, it will show that paper mail affords certain types of action, such as using the physical location of a letter to mark the stage at which some kind of "domestic task" has reached. Paper also supports a particular division of labour within the household, in which women exercise most control over domestic responsibilities. The chapter will argue that digital alternatives will need to mimic these features as well as offer added functionality if they are to achieve any kind of success in the home of the future. It will outline how this might be achieved as well as consider what are the technical and social limitations that will have to be thought about.

The next chapter reports a study that took seriously the need to understand social practices as part of the design process, in this case for the design of electronic programme guides for digital TV. Specifically it reports investigations of what will be called the natural rhythm of viewing. By this is meant the context in which TV viewing is undertaken and which gives it its particular meaning and value. Three types of viewing will be characterised. It will explain that each type also has a particular form of "viewer planning" and programme navigation. These involve a mixture of "common knowledge", now and next viewing habits, and occasional reference to newspaper listings. The chapter will explain how and why these programme navigation techniques should be designed into an Electronic Programme Guide, an EPG. A version of such an EPG will be presented, as well as alternatives currently in the marketplace. General lessons for the design of interactive entertainment systems will be drawn.

Most discussions of domestic Internet use focus on what families do online. The following chapter examines when and how family members organise the space of the home to enable PC use, as well as attend to the structuring of responsibility for the technology, which entails dealing with such things as teenage boys who tend to take on the job of "owning" the PC and the software on it. Here the issues are not solely related to the use of the PC and the Internet so much as a reflection of the use of scarce resources within the time and space of families generally. The families described in that chapter, from Boston and Pittsburgh, USA, are, in many respects normal and typical, though the treatment of family life in this chapter will have made them "anthropologically strange". The next chapter will report on the life circumstances of people who are not so much made strange by the analyst, but find themselves by dint of their medical history estranged from society and as a result find that the routine of living can be fraught if not dangerous. This chapter explores some of the difficult issues surrounding notions of "appropriate" design in "domestic" care settings. That is, it looks at the interaction between technologies, application domains, design methodologies and the challenges of informing design for households occupied by people with particular handicaps and needs. While in general this is hardly a novel concern, this particular focus arises as a consequence of digital technologies at long last maturing and transferring to this domain. When they do, they embrace various forms of "assistive" technologies and the design and provision of "smart" homes. This chapter presents some of the work of a recently initiated research project – "Care in the Digital Community" – whose objective is to facilitate the development of enabling technologies to assist care in the community for particular user groups with different support needs – sheltered housing residents and their staff. It considers the affordances of a variety of technological configurations including the use of virtual environments replicating real world situations and the use of handheld and wearable digital technology to provide support.

The baleful topic of this chapter ends the second section of the book. The third and last part, *the future home*, explores what it is like to live in smart homes. Commencing with discussions of how to make technologies fit into home life, chapter 10 attempts to identify what are the ways in which computing "disappears" from view. Such disappearing entails not only clever design but also providing users with the ability to meld the technology into their everyday lives. And, key to this, is the fact that such everyday lives are indeed, everyday, full of matters and concerns that are very ordinary, and to too many researchers, uninteresting, but it is just how these activities come to be so uninteresting that the authors of this chapter think that this makes them so relevant and important. Make them disappear and you have succeeded, is their maxim.

But what disappears may be surprising. The following chapter explores the use of online media, and finds that as online media will become more accessible to home consumers, so consumers will be less dependent on the computer as the gateway to information networks. Themes that come out relate to the creation and stability of audiences (i.e. relating to such questions as "who else is reading this newspaper?"), consumers' relation to "new" digital only features (such as hypertext links and archives etc.) and the "emotional relationship" consumers would appear to have to traditional newspapers. The chapter will explore why it is that consumers do not appear to develop the same with digital alternatives as they do with more traditional media, particularly the digital newspaper.

Smart homes have remained largely a myth rather than reality, and have rarely moved beyond interesting demonstrations at exhibitions and conferences. The next chapter will report what it is actually like to live in a smart house – albeit for short periods of time. Using data gathered through interviews and video, it will report on the experience of three families who occupied the Orange at Home house. Amongst other issues raised will be the peculiarities of using live subjects to explore radical and often intrusive technologies; another will be the social complexities of trying to link as many domestic technologies as possible. The limitations of the particular instantiation of a smart home will be discussed and reference made to the failures of smart house technologies in the past. Some of the successes and failures of this one from a technological and user perspective will also be discussed. Some remarks will be made about why certain types of technologies would appear to get more thorough design than others, and how this would seem to reflect certain naive assumptions about what is and is not necessary to sustain the desired patterns of home behaviour. These patterns are perceived by householders themselves as "normal", "reasonable and appropriate", though they are subject to gradual evolutionary change, sometimes driven by social factors and sometimes technological innovation.

The last chapter will not be about living with a smart house, as it is about who supplies the smart house, and here it should be obvious that such a house is more than just interactive systems or bricks and mortar. Building houses of whatever type involves a whole raft of suppliers and contractors, and each of these contribute wittingly or unwittingly to the ambience of the resulting product. To understand what a smart home is or what it can be is therefore also a case of understanding how the production of that house is undertaken. Though this is not about the living with a smart home, the book ends with this topic since it is a good way of tying up the knot that surrounds the issues that were introduced in the first substantive chapter of the book: the history of the smart home.

Through all of these chapters, the perspectives that have been covered are comprehensive, covering as they do the morality of the home, the economics of time and the psychology of space. They explore the use of smart home technologies from the perspective of wives, children and husbands, as well as the retired. But what the chapters don't do is report on one sector of society and one type of user, namely those who work at home. This may seem a strange absence since one of the driving ideas of smart home technologies is often thought to be to allow people to work at home.

There is nothing on these types of users in the book for two interrelated reasons. First, though historically there has been a great deal of research exploring what work at home might mean, most of it is rather dated. For one thing, far fewer people work at home now than they did in prior decades – in the UK, for example, the numbers are the lowest since 1950 – though there has been a shift in the type of persons who can work at home. If in the post-war years there was a lot of home work for women that consisted of piece work for manufacturing, now most of those who work at home undertake clerical or professional activities. Second, the theoretical concerns of this research were typically with the problem of how the home should defend itself against the invasion of work and, more particularly, capitalist forms of production.

Now, at the start of the 21st century, much of these concerns are irrelevant not only because the kinds of production in question are now no longer primary. It also has to do with how the reverse has happened. A great deal of contemporary research is showing that it is not work that goes home but home that goes to work. This is particularly the case with the web which allows people to undertake domestic activities like paying bills and monitoring accounts while at work. Mobile connectivity, too, allows the home to follow people about at work. Now, unfortunately, most of this research is yet to show itself in the public domain being confined to commercial research contexts. Some of my own research falls into this category.

However, the interesting thing about this is not that it will be appearing soon (as I am sure it will) but how this research is telling us more about things going on outside the home and less about what goes on in the home. For as I say this research is showing how the home is becoming the imperialistic domain of the 21st century, with the concerns of home invading every other space and moment; what needs to be investigated now is ways of managing this invasiveness and facilitating the related issues such as the increasingly complex transitions between home and work that people go through. In this sense, the topic for a future book will not be the smart home nor the smart workplace but the space that somehow exists in between: one facilitated by technology but populated by people who have to manage and evolve the concerns of each. In this book we will have elaborated on what those home concerns are; many other books deal with what work concerns might be. It is in this light that the absence of a concern for home work should be understood. And it is in this light that the merits of the book should be judged.

References

Castells, M (ed.) (1985) *High Technology, Space and Society*, California: Sage.

Castells, M (1998) *The Information Age: Economy, Society and Culture.* Vol. I. *The Rise of the Network Society*, Oxford: Blackwell.

Haddon, L (1992) "Explaining ICT Consumption: The Case of the Home Computer", in R Silverstone and E Hirsh (eds.), *Consuming Technologies: Media and Information in Domestic Spaces*, pp. 82–96, London: Routledge.

Harper, R (2000) "Domestic Design: An Introduction to the Research Issues Surrounding the Development and Design of Interactive Technologies in the Home", *Journal of Personal Technologies*, Special Issue on Domestic Computing, Vol. 4, No. 1, pp. 1–6.

Harper, R, Rouncefield, M and Randall, D (2000) *Retail Finance and organizational change: an ethnographic perspective*, London: Routledge.

Hindus, D (1999) "The Importance of Homes in Technology Research", *Co-operative Buildings Lecture Notes in Computer Science*, Vol. 1670, pp. 199–207.

Hughes, J, O'Brien, J and Rodden, T (1998) "Understanding Technology in Domestic Environments: Lessons for Cooperative Buildings", in *Proceedings of CoBuild '98*, Colorado: ACM Press.

Kling, R (1980) "Social Analyses of Computing: Theoretical Perspectives in Recent Empirical Research", *Computing Surveys*, Vol. 12, No. 1, pp. 61–110.

Nippert-Eng, C (1997) *Home and Work*, Chicago: Chicago University Press.

Venkatesh, A (1985) "A Conceptualisation of the Household/Technology Interaction", in *Advances in Consumer Research*, Vol. 12, pp. 151–55, Ann Arbour, MI: Association of Consumer Research.

Venkatesh, A (1996a) "An Ethnographic Study of Computing in the Home", Working Paper, University of California, Irvine.

Venkatesh, A (1996b) "Computers and Other Interactive Technologies for the Home", in *Communications of the ACM*, Vol. 39, No. 12, pp. 47–54.

Part 1
Conceptions of the Home

Smart Homes: Past, Present and Future

2

Frances K. Aldrich

2.1 Introduction

A "smart home" can be defined as a residence equipped with computing and information technology which anticipates and responds to the needs of the occupants, working to promote their comfort, convenience, security and entertainment through the management of technology within the home and connections to the world beyond.

The full-blown concept of the smart home is the acme of domestic technology we can envisage at present. The concept, at one time only encountered in science fiction, has moved closer to realisation over the last ten years. Although the gap between reality and fantasy is still wide, it is important that we start to give proper consideration to the implications this technology holds for the way we will live in our homes in the future.

To date, the limited amount of research into smart home that has been carried out has been primarily focused on the technical possibilities. As a social scientist myself, I am concerned that the personal and social consequences of smart home technology are largely being overlooked. (A notable exception is the work of Mynatt and colleagues at the Everyday Computing Lab at the Georgia Institute of Technology, e.g. Mynatt et al., 2001; Siio et al., 2002; Voida and Mynatt, 2002.) The home is a quintessential "human" place, with all the intricacies that entails. As I hope to persuade you in this chapter, the smart home is far too sensitive and important a sphere for social scientists to ignore any longer. If we take up the challenge which the smart home presents, we can make a significant contribution – the evolution of the technology itself will be shaped by our discussion and research.

With this chapter I am aiming to provide the motivation and the background for social scientists to become involved with the emerging phenomenon of the smart home. The chapter is divided into three sections looking at the past, the present and the future of the smart home – at the historical context which brought about the emergence of the "smart home" concept; its present status in terms of consumer take-up, current research projects and academic literature; and its future prospects, both commercially and as a potential area for social science research. The field of smart home research (and domestic technology in general) is in its infancy and relevant literature is sparse. I have therefore drawn together information from a range of disciplines and speculated where necessary, particularly regarding the future.

Finally, I should declare my own position – the social science of domestic technology is an area I find fascinating and I hope that I can communicate some of my enthusiasm in this chapter. Domestic technology has been referred to aptly as the "Cinderella" technology (Cockburn, 1997) – it simply does not get the creative urges of the male technology designers going (though hopefully the smart home may change that). Homes are still maintained largely by women's unpaid work but women have long been disenfranchised from the development of the domestic technology they use, playing little or no part in the design process which generally views them as passive consumers. It would be an oversimplification to see the whole neglect of domestic technology in male versus female terms but there is an element of truth in that view which, for me, adds to the drama and interest of this field of study. In fact, of course, research into domestic technology provides a greater opportunity to affect people's lifestyle and quality of life than research into many other technologies. In the developed world we are nearly all stakeholders in domestic technology and, to most of us, the home is a very important place.

2.2 Past History of the Smart Home

2.2.1 20th Century Domestic Technology: Seedbed of the Smart Home

The 20th century saw a dramatic revolution in domestic technology, a revolution which culminated at the close of the century with the emergence of the previously unimaginable concept of the "smart home".

At the beginning of the 20th century most of the available domestic technology would have been easily recognised and used by people from a hundred years earlier. By the end of the 20th century, however, domestic technology had changed beyond recognition. The first major impetus for change was the introduction of electricity into homes in the first quarter of the century. This provided a new source of clean, convenient power for appliances and spurred the introduction of novel equipment for the home. The second major impetus was the introduction of information technology in the last quarter of the century. This opened up possibilities for exchanging information between people, appliances, systems and networks in and beyond the home, possibilities which are still being explored.

In the following brief historical review (with due acknowledgement to Gann et al., 1999), I have attempted to capture the escalating pace and dramatic nature of developments in domestic technology across the 20th century – changes which prepared the "seedbed" for the emergence of the smart home.

- *1915–20:* During the early part of the century the emerging middle-classes were experiencing a shortage of domestic servants (Forty, 1986). In line with this labour shortage, electrically powered machines such as vacuum cleaners, food processors, and sewing machines were introduced into the home for the first time. The advertising angle was that with the help of technology, one person alone (inevitably a woman) could manage all the household chores and still have time for leisure activities (Hardyment, 1988). Advertisements used phrases such as "spring cleaning with electricity", "no longer tied down by housework", and "automatically gives you time to do those things you want to do" (Gann et al., 1999). Mains electricity was not yet widespread, however, and so for most housewives such images remained a high-tech fantasy.

- *1920–40:* By 1940 the proportion of households in UK with mains electricity had risen to around 65 per cent. Many homes still only had electricity for lighting, however, while others had just one 5 amp socket. People sometimes declined to pay the additional cost to have a socket fitted as they were unable to envisage any use for one. Within the home the emphasis switched from production to consumption, with advertisers attempting to understand and appeal to the psychology of the housewife, "Mrs Consumer" (Frederick, 1929). It is an irony that the introduction of new domestic technology actually resulted in women spending more time on housework than ever before, because standards rose – washing machines led to clothes being washed more often (Cowan, 1983), and vacuum cleaners led to floors being cleaned more frequently (Hardyment, 1988).

- *1940–45:* During the Second World War, government propaganda portrayed women as technically competent and stressed the valuable role they could play in taking over traditionally male jobs in manufacturing and industry, freeing men to go into the armed forces. Women grew accustomed to working outside the home and (as illustrated by the well-known film *Rosie the Riveter*) many became technically proficient and enjoyed these new roles. Through working in these comparatively well-paid jobs women also came to value their labour in financial terms. These factors helped to pave the way for the uptake of domestic technology after the war.

- *1945–1959:* After the Second World War, in order to free jobs for men returning to civilian life, government propaganda switched to persuading women that their place now was back in the home. Advertisements of the time show women in the home, waving husbands and children off for the day, and then turning their attention to the daily domestic fight with "germs rather than Germans", as it has been expressed. Home design started to reflect new ways of living alongside modern technology. For example new styles of kitchen emerged to accommodate the refrigerators, electric cookers, and washing machines that were starting to penetrate the domestic market. The concept of the "television lounge" was introduced, the sale of televisions having increased massively prior to the Coronation.

- *1960s/70s:* The 1950s ideal of the stay-at-home-housewife was overturned during the "swinging 60s". With the contraceptive pill and greater choice about whether and when to have children, more women started to go out to work. Numerous labour-saving devices became common in the home, including kettles, toasters, cookers, coffee and tea makers, food processors, hair dryers, electric razors, washing machines, sewing machines, vacuum cleaners, and irons. Other technology became commonplace in the home, for example central heating and thermostats.

- *1980s/90s:* By the beginning of the 1980s, almost three-quarters of households in England and Wales had colour television, and by the end of the 80s half also had video recorders (Bowden and Offer, 1994). Microwave ovens, freezers and tumble dryers also became increasingly common during this period which, in addition, saw the introduction of cordless and mobile phones for domestic use. A host of new home entertainment technologies became available and started to penetrate the domestic market – cable TV, DVD, the playstation, and the multimedia PC. The migration of the PC from workplace to home is particularly significant because it opened up the possibility of teleworking, blurring the distinction between home and work. Furthermore, by allowing access to the Internet, the PC connected the home to a host of new services such as banking, shopping and information, services which are still evolving.

From this brief review it is apparent that different forms of domestic technology have been adopted at different rates. As this may have implications for the uptake of smart home technology in the future, it is worth considering an important distinction known to have influenced diffusion rates in the past – the distinction between "time-saving" goods and "time-using" goods (Bowden and Offer, 1994). "Time-saving" goods are those which can potentially increase discretionary time by reducing the time needed to carry out a task, for example washing machines. "Time-using" goods are those which occupy discretionary time and improve its perceived quality, for example television. The diffusion of "time-saving" goods such as the vacuum cleaner, refrigerator and washing machine took several decades and was clearly related to household income. In contrast, radio, television and video (all "time-using" goods) reached equivalent levels of diffusion within a few years and showed much less relationship to household income.

A further point worth noting with regard to future smart home technologies is that "time-saving" and "time-using" goods may compete for the amount of time allocated to their use. The increasing amount of time people spent watching television was found by reducing time spent on housework, which had previously risen steadily for years as standards of hygiene became higher (Bowden and Offer, 1994). There may be a historical lesson to learn here.

In summary then, the 20th century saw an increasing pace of change in domestic technology, a readiness to adopt "time-using" technologies in particular and, by the end of the century, many homes linked via the PC to information and services beyond the home. This then was the seedbed in which the concept of the smart home developed.

2.2.2 Emergence of the "Smart Home" Concept

Advanced home control systems go by several names, including smart home, home automation and integrated home systems. By any name, these systems conveniently control home electronics and appliances including audio/video, home office, telecommunications, intercom, security, lighting, HVAC, and lawn sprinklers. Control systems can also provide information – residents can find out how much electricity they've used on specific appliances or systems, and utilities can read meters remotely. The systems can be accessed from remote locations by phone or computer, allowing residents to turn on the heat, for example, on their way home from work (*Home Energy Magazine Online* May/June 1998).

Interest in "wiring" homes for increased functionality, as described above, dates back at least to the 1960s. At this time it was largely the province of home hobbyists, however, and most other people would have considered the description above to be science fiction.

By 1984, however, commercial interest in home automation had grown sufficiently for the National Association of Home Builders in the USA to form a special interest group called "Smart House" to push for the inclusion of the necessary technology into the design of new homes. Interest came principally from the fields of building, electronics, architecture, energy conservation, and telecommunications. Social scientists showed no real interest in the smart home concept.

Since the 1980s, manufacturers of consumer electronics and electrical equipment have been developing digital systems and components suitable for use in domestic buildings. Important developments have included the replacement of electromechanical switching with digital switching, and of traditional twisted pair and coaxial cables by optical fibres. Other enabling developments are new communication networks (e.g. ISDN, Internet) which allow two-way communication, and new end devices (e.g. web TV, video phones) (Barlow and Gann 1998).

During the 1990s the concept of the smart home entered popular culture for the first time. No longer the province solely of sci-fi buffs and electronics hobbyists, smart home articles began to appear in life-style magazines such as *Boys' Life*, *Vanity Fair* and *House Beautiful*. The BBC recently ran a television documentary series entitled *DreamHouse*, which followed a family living in an experimental smart home for six weeks, giving viewers some idea of what it might be like to live in such a house. However, despite greater public awareness of the smart home concept, the extent to which people at the end of the 20th century were ready to welcome such technology into their own homes was uncertain. Popular media suggested some apprehension, particularly over issues of retaining control over the technology. Although this was not a new concern (see, for example, the film *Demon Seed* released in 1977), it remained a relevant theme as indicated by the film *Dream House* (1998) in which a malevolent smart home takes control over its occupants. The enduring unease about smart home technology has been neatly expressed by Gold (quoted by Gibbs, 2000), with his question "How smart does the bed in your house have to be before you are afraid to go to sleep at night?"

2.3 Present Status of the Smart Home

2.3.1 Consumer Take-up

Although the concept of the "smart house" was well established by the end of the 1990s, to date only a small number of expensive "smart homes" have been built and sold on the commercial market, in contrast to the rapid diffusion envisaged.

Gann et al. (1999) suggest a number of reasons for the slow uptake of smart home technology. The principal barriers to uptake they identify are that:

- The initial investment required from the consumer is high, restricting the market to the middle and upper income brackets, and potential buyers must first be convinced of the benefits they will derive.
- In Europe, dependence on old housing stock means manufacturers must find solutions for "retrofitting" existing housing, which is more expensive than networking a home at the time it is built (Barlow and Gann, 1998).
- Because of the lack of a common protocol, the smart homes industry in Europe has tended to focus on simple on-off switching systems (e.g. remote control switching) for single applications, which require no additional network installation.
- Suppliers have adopted a narrow "technology push" approach and paid too little attention to understanding the needs of users. Consumers want systems which will help them with managing everyday tasks, offer labour saving and task simplification, ease of operation, remote control and cost reduction (Meyer and Schulze, 1996). There is a gap

between consumer requirements and the products currently available. In particular, Meyer and Schulze suggest, suppliers need to win the acceptance of women, who still remain responsible for the bulk of domestic tasks.

- Suppliers have done little to evaluate the usability of their products. Barlow (1997) points out that this is not a simple task, however, because of the diversity of the user population, variation in the context of use, prior training necessary, and the challenge of investigating products not yet in existence.

The smart homes industry must satisfy a number of criteria before consumers will be motivated to buy its products, Barlow and Gann (1998) suggest. First, the industry must develop solutions which meet real user needs. Secondly, the solutions must operate at three levels – as *generic technologies,* providing basic, compatible "building blocks" for *context-specific systems* (which can be adapted to a wide variety of dwellings) and for *personalised systems* (tailored to the requirements of specific individuals and households). Thirdly, the solutions must offer functionality; ease of use; affordability; reliability and maintainability; flexibility, adaptability and upgradability; and replicability and ease of installation.

Interestingly, Gann et al. (1999) have pointed out a number of parallels between the present market for smart home systems and the early market for electrical appliances. Before demand for electrical appliances took off, a number of preconditions had to be met, including a cheap supply of electricity, cheap and reliable appliances, and the installation of a distribution and wiring system. Initially, most homes with electricity had only one 5 amp socket and people were not convinced of the value of having more – those homes which had more than two sockets were usually newly built houses in the upper price range (Forty, 1986). In a further parallel with smart homes, potential consumers showed apprehension about the risks of electricity; fears had to be allayed and acceptance gained by demonstrating its advantages.

2.3.2 Experimental Projects

Despite the fact that consumer uptake has disappointed the pundits, there are now quite a few demonstration smart homes in existence. Many of these are simply commercial showcases with no research agenda. However, there are also a number of commercial projects actively exploring the possibilities offered by technology associated with the smart home, for example utility companies seeking to control domestic energy consumption remotely. These investigations are interesting because they can be regarded as experiments in the "real world". Their drawback is that, where they are not academia-led and involve no social scientists, evaluation from a user-centred perspective may not be thorough. Furthermore, the findings may never enter the public domain.

For these reasons I will limit the review of experimental smart homes below to academic projects exploring the implications of the technology. This leaves a small field in which the following projects are most notable:

The Adaptive House (University of Colorado)

The aim of the Adaptive House experiment is to explore the concept of a home which programs itself, freeing the inhabitants from the need to carry out this task. The researchers point out that the software for an automated home must be programmed for a particular family and home, and updated in line with changes in their lifestyle. Given that many people find it difficult enough to program their video recorders, programming a smart home will be beyond their interest and capability, and hiring a professional to do the job would be costly and inconvenient.

The prototype system is installed in the home of one of the researchers and controls room temperature, water heat, ventilation and lighting. The home is equipped with sensors which monitor temperature, light levels, sound, the opening of windows and so on, as well as control devices for heating, lighting, fans, etc. The system monitors actions taken by the residents, such as turning on a certain configuration of lights, or turning up the thermostat, and looks for patterns in the environment which reliably predict these actions. A neural network learns these patterns and the system then performs the learned actions automatically. (See Mozer, 1998; and http://www.cs.colorado.edu/~mozer/house/)

ComHOME (The Interactive Institute, Sweden)

The ComHOME project is described by the researchers as "a full-scale model constructed of a number of scenario-like room set-ups" (Junestrand and Tollmar, 1999). The apartment is equipped with technologies such as sensors, voice control and voice-mediated communication. In this context researchers are investigating different spheres of home-based activity, for example communication, distance work and social activities, and exploring the impact which technology may have on them. (See Junestrand and Tollmar, 1999; also http://cid.nada.kth.se/pdf/cid_61.pdf).

House_n (Massachusetts Institute of Technology)

House_n is a collaborative, multi-disciplinary project led by the Department of Architecture. The overall aims include creating environments which suit people of all ages; creating customisable environments; developing algorithms to interpret sensor data to detect what people are doing; exploring the impact of technology on traditional learning environments; inventing interfaces and components that conserve resources; and exploring the impact of home delivery of products and services. There are plans for a "living lab" house but in the meantime a large workshop room is being equipped as a prototype.

In the workshop it will be possible to display digital information on almost any surface, with other surfaces allowing easy user input via touch or special devices. A partition allows division of the floor space for living and sleeping, and provides a "medical nook" for the receipt and analysis of medical information. It is planned to explore a variety of home activities within this context, using an "active counter" that can be used for kitchen tasks, work tasks, and eating; an "active table" with digital surface that can be moved around within the environment; "video walls"; and floors that can have video projected onto them.

Researchers in the Media Lab at MIT are meanwhile exploring a vision of the kitchen of the future as a digitally connected, self-aware environment with memory of its actions. Concepts and prototypes include a variety of intelligent appliances as well as an intelligent work surface.

(For House_n see http://architecture.mit.edu/house_n/. For the Media Lab kitchen projects – e.g. Counter Intelligence, CounterActive and Kitchen Sync – see http://gn.www.media.mit.edu/pia/Research/index. html).

The Aware Home (Georgia Institute of Technology)

Most of the research by the Aware Home Research Initiative takes place in the Broadband Institute's Residential Lab – a suburban house equipped with high-speed internal and external connections, cameras and microphones, a house-wide wireless net allowing communication between cordless devices, and a radio-locating system for tracking tagged objects. At the time of writing the house had not been lived in. The Aware Home project is arguably the most well-advanced of the smart home research projects, involving researchers from the Broadband Institute, the Everyday Computing Lab, and the Future Computing Environments Group.

The over-arching theme which has been adopted is to use the technology to help maintain older people in their own homes for as long as possible. There are two focuses to the research: first, issues and possibilities concerned with making the house aware of the whereabouts and activities of its occupants at all times; and secondly, the implications of maintaining continuous connectivity to the electronic world, particularly as a means to "reunite the nuclear family of the 21st century". There are a wide variety of concepts and projects associated with the Aware Home and in different stages of development. Examples include: software which automatically constructs family albums from video pictures collected in the house; an intercom system which uses voice recognition to allow people to speak to one another by saying their name; software that telephones a person when their photograph is spoken to (after first checking they are awake); electronic tagging of easily mislaid items such as keys and remote controls; reminders from the house about appointments, medication, etc., through subtle images and sounds; a "smart floor" system which identifies and tracks people by their footsteps; digital portraits incorporating iconic data representing the physical and social well-being of the Aware Home occupant; and a smart environment (the kitchen in particular is mentioned) that records contextual information alongside a record of everyday activities to help people resume interrupted activities. (See Kidd et al., 1999; Gibbs, 2000; http://www.cc.gatech.edu/fce/ahri/ and http://www.broadband.gatech.edu/facilities/resident/resident.htm).

Summary

It is important to emphasise that among the experimental smart homes reviewed above only the Adaptive Home is occupied. This obviously limits the scope of the conclusions which can be drawn from the current research – much of the complexity of the home environment only emerges in the interplay of activities and relationships between household residents.

2.3.3 Academic Literature

Although the concept of the smart home is now well established and a number of research projects are underway, as a field of academic research the smart home is still in its infancy. This is perhaps not surprising because, as we saw earlier, domestic technology in general has been neglected by academics, despite the enormous changes seen over the last century. Reasons for this neglect have been examined by an established body of feminist research (see, for example, Wajcman, 1991). Chief among the reasons identified are lack of financial motivation to increase productivity in domestic work; little involvement of the technology users in the design process; product designers' view of domestic technology as unexciting; and a continued focus on stand-alone appliances, often for marketing reasons.

In what will hopefully become a landmark paper, "The Importance of Homes in Technology Research", Hindus (1999) calls for more academic interest in domestic technology on the grounds that it is too economically important to ignore and that research has the potential to improve everyday life for millions of users. She points out that although information technology may have migrated from the workplace to the home, research specific to the home is still needed because workplace findings cannot easily be generalised to the home context. She points out three reasons for this. First, "homes are not workplaces" – unlike workplaces they are not designed to accommodate technology, they are not networked, nor do they have the benefit of professional planning, installation and maintenance of technology and infrastructure. Households also include elderly people, children, babies and pets, as well as working age adults. Secondly, *"consumers are not knowledge workers"* – motivations, concerns, resources and decisions are different in the home. Whereas workplace purchasing decisions are determined by concern with productivity, householders are interested in aesthetics, fashion and self-image. Thirdly, *"families are not organisations"* – they are not structured in the way that corporate organisations are structured, and decision-making and value-setting are quite different.

A few papers on smart homes are now beginning to emerge, however, generally in association with one of the experimental projects outlined earlier. However, these usually approach smart homes from the technical point of view (e.g. Mozer, 1998). A paper which is unusual in approaching smart homes from the perspective of social science is that by Berg (1994). In her paper she argues that the smart home is a "gendered socio-technical construction" developed in line with the interests of its male designers. She focuses on housework which she describes as "mainly women's unpaid work, compris[ing] the most repetitive and time-consuming tasks in the household – cooking, washing, cleaning, tidying, mending". She interviewed the designers of a number of experimental smart homes, asking how they thought technology might help and found that the designers "manifest[ed] neither interest in nor knowledge of housework. The home is acknowledged as an important area of everyday life, yet the work that sustains it is rendered invisible." She observed that

> the men (and it is men) producing prototypes of the intelligent house of the future and designing its key technologies have failed to visualise in any detail the user/consumer of their innovation. In so far as they have one in mind, it is someone in their own image. They have ignored the fact that the home is a place of work (women's housework) and overlook women, whose domain they are in effect transforming, as a target consumer group (p. 176).

She concludes by criticising the smart home as a typical case of "technology push" rather than "consumer pull", motivated principally by what is technically possible rather than what is desirable.

Publications of relevance to smart homes, and to domestic technology generally, are currently dispersed across a wide range of academic disciplines. Those looking to the literature for guidance must pick their way across a fragmented area, gleaning what they can where they can. This is an unsatisfactory situation as the potential offered by smart home technology can only be realised through a proper understanding of the complex social context in which it will be used. Like Hindus (1999) and others, I hope technology in the home will receive more serious attention in the 21st century.

2.4 Future Prospects for the Smart Home

The commercial outlook for smart homes is still a matter for speculation. A recent report by Barlow and Venables (2001) discusses this topic in some detail, and from a more technical standpoint than would be appropriate here. Barlow and Venables also consider the issue in Chapter 13 of this book. However, the present chapter would be incomplete without some discussion of the smart home's commercial future, and particularly the scope for social scientists to influence the course of developments. We shall therefore look briefly at likely changes to the markets, main players and barriers to take-up.

2.4.1 Markets

The distinction between the "niche market" and "generic market" looks set to continue.

The niche market caters for the needs of special groups such as elderly and disabled people. By definition, therefore, it is a smaller market but one with the potential to take off rapidly if the cost-saving benefits of providing health care and practical support through home technology can be established. Attempts to do this are underway (e.g. Tang et al., 2000).

The generic market refers to the population as a whole and is therefore potentially a huge market in comparison. However, as yet no smart home technology has had the "must have" quality which led to the rapid diffusion of television, for example, with take-up largely independent of household income. As we saw earlier, "time-using" technology has been adopted more swiftly than "time-saving" technology in the past. This suggests that inroads into the generic smart house market are most likely to be made by those technologies which add to the perceived quality of discretionary time.

Social scientists have an important role to play in developing a better understanding of the niche and generic markets for smart home technology. Through sensitive exploration of user needs and values, we can help to ensure that technological developments will offer genuine benefits, in contrast to the "push" technology which dominates the market at present.

2.4.2 Main Players

There is some indication that the main players in the commercial market for smart home technology are changing. Historically, it has been the electrical equipment suppliers – manufacturers of switches, sockets and distribution boards for example – who have played the leading role in developing the market. Now, however, there is established interest from consumer electronics manufacturers such as Nokia, Sony and National Panasonic, and ranges of so-called "smart" appliances are being marketed directly to the household consumer.

Service providers are also taking an increasing lead in exploring the market for smart home technology. A number of electricity companies in Europe and the States are looking into the provision of services which allow householders to control their heating, lighting, security equipment and other appliances remotely, by telephone or computer, with signals sent over the existing electrical wiring in their homes (Handford, 2002). These companies also have an interest in using the same technology for energy conservation and management of demand.

Again social scientists have an important role to play. The possibility exists for the service companies themselves to take some control over the technology within people's homes. Will this be acceptable to people? Will the gains in terms of savings outweigh the costs in terms of loss of privacy and autonomy? Commercial companies are already starting to explore whether consumers are prepared to have this kind of relationship.

2.4.3 Obstacles to Consumer Take-up

Earlier on we looked at a number of current obstacles to consumer take-up of smart home technology. We will review these now and consider whether they are likely to change in the near future:

- *Dependence on old housing stock* – there is no prospect that this will change in Europe and manufacturers must continue to look for ways of equipping houses retrospectively.
- *Lack of a common protocol* – this is becoming less of an obstacle than previously because there are now home boxes which can cope with different protocols.
- *High initial investment from the consumer* – cost remains relatively high and potential buyers are yet to be convinced of the benefits. A likely development is the evolution of a more modular system of smart home technology which people can acquire in stages.
- *Little usability evaluation by suppliers* – it remains the case that insufficient attention is paid to usability of smart home technology and it seems unlikely that attitudes are about to change.
- *"Technology push" by suppliers* – suppliers are still paying too little attention to the needs of users and trying to market products and services for which there is no demand.

Some of these obstacles look set to remain while others are beginning to shift. The skills of social scientists are particularly needed to overcome the last two obstacles on the list. First, whether or not social scientists choose to become involved in the evaluation of usability themselves, they should make their voices heard in pushing for it to happen – it is a vital part of the design process if the technology is to become acceptable to the user. Secondly, social scientists are well placed to play a major part in developing a proper understanding of user needs and it is this which will overcome the present "push" of unwanted technology to uninterested consumers.

2.4.4 Academic Literature: What to Watch

So few papers on smart home technology are available in the academic literature, one can only speculate which fields to watch for future developments. The obvious usability issues in the design of appliances, interfaces and systems for the smart home invite the involvement of human factor researchers. However, the context of use which the home provides is complex, social and cultural, suggesting the need for sociologists, anthropologists, ethnographers and social psychologists to contribute too. A number of people have pointed to multidisciplinarity as the way forward for the design of information technologies in general. For example, Norman (1998) identified the following sets of skills as important within the ideal research and design team: anthropology, sociology, cognitive science, experimental psychology, human-computer interaction, architecture, industrial design and art.

For the time being we can expect to see publications of some relevance to the smart home scattered across the literature of a number of academic disciplines. However, there are signs of growing interest in the home as a context of use of interactive technology (e.g., O'Brien et al., 1996; Scholtz et al., 1996; Tollmar and Junestrand, 1998; and this volume) and it is realistic to hope that a dedicated journal may emerge as a focus for this interest within the next decade.

Until then, it is important not to overlook the World Wide Web as a means of keeping abreast of developments concerning the smart home – a means which offers the advantage of providing some information about commercial research ventures as well.

2.4.5 Foundations for Future Research: Relevant Categories and Constructs

A number of categories and constructs concerning smart homes, and domestic technology in general, have emerged from the literature so far. The distinctions concern both the technology and its users, and range in level of analysis from classifying homes as a whole to considering occupants' behaviour at a minute-by-minute level. These categories and concepts are valuable because they prompt one to shift perspective, so gaining different insights into homes and the use of technology within them. They may well provide the foundations for future research, so I review them here under the following headings: smart homes, households, activities, and technologies.

Smart Homes

There are various ways of conceptualising the organisation of elements which make up a smart home. These range from focusing closely on the technology, to a looser biological metaphor for the management of information. Five examples from the literature are given below and later I propose a sixth.

- Dard (1996) focuses on the flow of information about activities and resources within the home. He classifies three information flows: *human flows* (e.g. supervising private and shared spaces); *energy flows* (e.g. monitoring energy consumption); and *information flows* (e.g. managing transmission and reception of messages).

- Barlow and Gann (1998) focus on the technology. They distinguish three levels of technology: *generic technologies* which provide compatible building blocks for more elaborate systems; *context-specific systems* adapted to a variety of dwellings; and *personalised systems* tailored to individual and household requirements. The authors also consider the level of automation which the technology permits, distinguishing between *fixed* applications, *programmable* applications, and *automated* applications.

- Jedamzik (2001) focuses on both the control and the information which is available to the user, and proposes that a smart house has four components: *user interface*; *technical field* (controlling light, heat, climate, water); *field of information* (where the house serves as a knowledge base, e.g. health, household accounting, scheduling); and *service field* (connecting to external services, e.g. financial, legal, commercial, educational).

- Gann et al. (1999) also focus on the functionality available to the user, distinguishing two forms of smart home. In the first, the emphasis is on *intelligent appliances* . This is the more traditional approach to home automation. The other involves *interactive computing within and beyond the home* which has come to the fore more recently.

- Another approach, focusing on control of the home environment, adopts the biological metaphor of the sympathetic and para-sympathetic nervous system in animals. This approach highlights that while certain aspects of home control require the user to exercise conscious thought and deliberate action, it may be desirable to monitor and control other aspects automatically (e.g. lighting and temperature), freeing the user for other tasks.

This variety of perspectives is useful for the different ways it gives us of looking at the smart home, its technology, and its users.

Households

As well as considering the smart home as a technological entity, it may be useful to categorise the residents along a number of dimensions, with the aim of understanding how various family set-ups or household types make different use of technology available for the home.

Meyer and Schulze (1996) have already made an attempt to do this for the purpose of predicting uptake of smart home technology. They suggested that uptake will depend in particular on size and composition of the household, the division of labour, and stage in the family lifecycle. They proposed that the households with the most to gain from adopting smart home systems are those in which both partners are working; highly mobile single-person households; and households with elderly or disabled people.

Of course there may be other dimensions of relevance, for example the geographical spread of family and friends, and whether the household is located in a rural, suburban or urban setting.

Activities

Analysing the activities which take place within a home may also be a productive means of considering the use of domestic technology. There are several possible frameworks to use in conducting such an analysis:

- *Spatial framework* – the home may be considered in terms of spatial "zones" within the home, which may or may not map onto particular rooms, but which are differentiated by the type of activity that is carried out in that location. For example, an ethnographic study by Mateas et al. (1996) showed space in the home is not of equal significance but shows behavioural clusters such as "Work space" and "Hang-out space" (often the kitchen, where families spend much of their time).

- *Temporal framework* – the activity within a home may also be considered from a time perspective. For example Mateas et al. (1996) looked at the way time was structured during the day and found the idea of large blocks of free time was a myth. Instead the day consisted of many small blocks of time, each constrained to varying degrees by a variety of factors. An alternative to the "time blocks" approach is to look for temporal patterns in activities, an approach already well established in the study of conversation where it is referred to as "sequential organisation", e.g. Sacks (1992).

- *Goal-oriented framework* – some activities may not be captured effectively within either a spatial or temporal framework. These are complex sequences of activity which are neither spatially nor temporally contiguous. An example is packing to go on holiday, which might involve planning what to take; washing and ironing clothes; ordering foreign currency; finding a suitcase; buying a novel for the journey, etc. – tasks which may take place over several days and across several locations, not all of them within the home. A complex sequence of activity like this could be broken down into a series of shorter component activities which do fit into a spatial or temporal framework. However, it is important to maintain a sense of the overarching purpose of the component activities, so any framework should incorporate goal-orientation.

- *Communication framework* – finally it is worth mentioning that in deriving a model of household activity from their ethnographic data, Mateas et al. (1996) gave a special status to communicative activities and considered these separately. They found, for example, that most communicative activity took place between family members in the same location (supplemented by contact with remote family and friends), and was highly valued within the family system.

Technologies

In addition to investigating types of households and activities, it may be helpful to categorise the technology itself with a view to looking for patterns of behaviour and technology use which map onto these categories.

Let us consider the well-established distinction between "white goods" and "brown goods" as an example (e.g. Cockburn and Ormrod, 1993). These two categories of domestic technology are contrasted in Table 2.1, illustrating some of the patterns in behaviour and technology use that emerge from this approach.

Summary

The categories and constructs we have reviewed above (relating to types of home, household, activity and technology) are valuable for the different insights each provides. This is useful groundwork but we need much more research if we are to gain a proper understanding of the way technology is and might be used in the home.

2.4.6 When is a "Smart Home" not a Smart Home?

When is a "smart home" not a smart home? Or, put another way, when *is* a "smart home" a smart home? These questions may sound trite but I raise them with serious intent and suggest that attempting to answer them will pay dividends in terms of improving the clarity of thought and vision surrounding the smart home concept.

As yet there are no industry standards governing use of the term "smart home" and it is applied very loosely – to anything from a home with a closed-circuit television security system to a ground-breaking

Table 2.1. White goods and brown goods

	White goods	Brown goods
Example	Washing machine, cooker, vacuum cleaner, microwave	Hi-fi, TV, VCR, PC, camcorder, cable, games console, Internet
Function	Domestic work	Leisure
Effect on time use	Time-saving	Time-using
Underlying technology	Mechanical	Electronic
	Electrical	Computer
Orientation	Self-contained	Bring "outside in"
Designers' attitude	"Pedestrian"	"Leading-edge"
Exposure in home	Behind the scenes	On show
Consumer uptake	Push	Pull
Gender stereotype	Female	Male
Workplace findings	Not relevant	Some limited relevance where technology "domesticated"

demonstration house. Back in 1989, Forester sounded this cautionary note: "a combination of home computers, consumer electrical goods, videotext services, and home security systems, even in a 'smart house', wired with heating and lighting sensors . . . hardly adds up to a revolution in ways of living". I agree; even now much of what is presented as radical and new proves to be unexciting on close scrutiny.

My personal view is that the smart home does hold the potential for a paradigmatic shift in the way people live with technology at home. As things stand, however, it is hard to analyse that potential. None of the distinctions between smart homes reviewed above seems to capture it, and the frequently exaggerated claims for so-called "smart homes" simply muddy the waters. Trying to analyse what is genuinely new and different in the opportunities offered by smart home technology would be a good starting point for more insightful design and further breakthroughs in the future. Technologists can only take things so far – social scientists have an important role to play too.

To start the ball rolling I present a classification between smart homes which is finer-grained than those reviewed above. My aim is to capture the scope for a paradigmatic shift in the way we live with domestic technology, although the reality is not yet with us. My starting point is the distinction drawn by Gann et al. (1999) between homes which simply contain smart appliances, and those which allow interactive computing in and beyond the home. Maintaining this focus on the functionality available to the user, I propose five hierarchical classes of smart home:

1. *Homes which contain intelligent objects* – homes contain single, stand-alone appliances and objects which function in an intelligent manner.

2. *Homes which contain intelligent, communicating objects* – homes contain appliances and objects which function intelligently in their own right and which also exchange information between one another to increase functionality.

3. *Connected homes* – homes have internal and external networks, allowing interactive and remote control of systems, as well as access to services and information, both from within and beyond the home.

4. *Learning homes* – patterns of activity in the homes are recorded and the accumulated data are used to anticipate users' needs and to control the technology accordingly. (See, for example, the Adaptive House which learns heating and lighting usage patterns, Mozer, 1998.)

5. *Attentive homes* – the activity and location of people and objects within the homes are constantly registered, and this information is used to control technology in anticipation of the occupants' needs. (See, for example, the Aware Home, Kidd et al., 1999.)

This classification of smart homes highlights different levels of communication of information within and beyond the home; distinguishes systems which can learn from those which cannot; and differentiates homes which maintain constant awareness of occupants and objects from those which do not. The classification is also hierarchical: from the users' perspective, each level promises some increase in functionality; from the technical perspective, each level generally depends on the systems for the previous level being in place.[1]

Within this classification of smart homes, the issue of control over appliances and systems in the home emerges as a strong underlying theme. Moving up the hierarchical classification, the control systems involved range from the simplest switching mechanisms (which respond only to direct on-off signals) to highly complex systems (capable of interpreting and responding to complex external stimuli such as people and their activities). The opportunity for occupants to delegate control to the technology increases correspondingly. It is this handing over of control that increases potential functionality in the smart home – the house itself can be empowered to perform a greater range of tasks relating to the occupants' comfort, convenience, security and entertainment.

If a paradigmatic shift in the way we live with domestic technology is going to occur, I suggest that it is the implementation of the fifth level of smart home, the Attentive Home, which will bring the shift about. The Attentive Home, with its potential for flexibility, proactivity and responsiveness to the user, appears to offer the possibility of a home environment qualitatively different to any we have seen before.

---

[1] The order of the last two categories is debatable; however, I argue that monitoring household activity and anticipating needs in real-time is a technically more demanding task than learning general patterns of use across time, and that meeting this challenge would offer greater functionality for the user.

2.5 Conclusion

To summarise, we looked at the emergence of the smart home concept towards the end of the 20th century, placing it in historical context by reviewing the rapid developments in domestic technology brought about first by electricity, and secondly by information technology.

We examined current obstacles to smart home technology diffusion, reviewed the principal academic projects investigating smart home issues, and considered what the available academic literature has to offer and reasons for its paucity.

We looked at possible future developments in terms of commercial players and markets, and likely academic stakeholders. We also considered categories and constructs relating to homes, households, activities and technology which may provide a useful foundation for future research in this area. Finally we considered the question "When is a 'smart home' not a smart home?", the reasons for asking this question, and an extended classification of smart homes which starts to address the issue.

Considering this and other issues related to technology in the home is undoubtedly made harder by the lack of a body of evidence on the design and use of technology in the home setting. It is necessary to draw on literature from across a range of disciplines to piece together a more complete picture, and many gaps remain. As Hindus (1999) pointed out, there is not yet a "critical mass" of interest in this area from academic researchers, or for that matter, industry. Hopefully her paper will mark a turning point and domestic technology will become established as a field of study in its own right.

In smart home research in the meantime, issues for investigation, methodologies, research paradigms, and frameworks for analysis must all be decided on with little guidance from the literature. Research must necessarily be pioneering in approach and holds the potential for considerable impact, not only shaping the course of future research, but determining the nature of the very homes we will live in. This is an interesting field of research for social scientists and an exciting time to become involved.

Acknowledgements

My sincere thanks to Blay Whitby, Paul Martin and Tim Venables for helpful discussions before I started this chapter, to Tim Venables for commenting on the initial draft, and to Abi Sellen for getting me involved with smart homes in the first place. My thanks also to Richard Harper, Dave Randall, Lynne Hamill, John Strain and Peter Morris, whose variety of perspectives from different disciplines have led to stimulating group discussions along the way.

References

Barlow, J (1997) "Smart Homes Project. User Needs Analysis", Report to the Joseph Rowntree Foundation. Mimeo.

Barlow, J and Gann, D (1998) "A Changing Sense of Place: Are Integrated IT Systems Reshaping the Home?", paper presented to the Technological Futures, Urban Futures Conference, Durham, 23–24 April.

Barlow, J and Venables, T (2001) "The Evolving User Environment", in *Future Bottlenecks in the Information Society*. Report to the EU Parliament from the European Science and Technology Observatory.

Berg, C (1994) "A Gendered Socio-technical Construction: The Smart House", in C Cockburn and R Furst-Dilic (eds.), *Bringing Technology Home: Gender and Technology in a Changing Europe*, Buckingham: Open University Press.

Bowden, S and Offer, A (1994) "Household Appliances and the Use of Time: The United States and Britain Since the 1920s", *Economic History Review*, Vol. XLVII, No. 4, pp. 725–48.

Cockburn, C (1997) "Domestic Technologies: Cinderella and the Engineers", *Women's Studies International Forum*, Vol. 20, No. 3, pp. 361–71.

Cockburn, C and Ormrod, S (1993) *Gender and Technology in the Making*, London: Sage.

Cowan, RS (1983) *More Work for Mother: The Ironies of Household Technology from the Open Hearth to the Microwave*, New York: Basic Books.

Dard, P (1996) "Dilemmas of Telesurveillance in Housing", paper presented at the ENHR Housing Conference, Helsingor, August.

Forester, T (1989) "The Myth of the Electronic cottage", in T Forester (ed.), *Computers in the Human Context: Information Technology, Productivity and People*, Oxford: Blackwell.

Forty, A (1986) "Objects of Desire: Design and Society 1750–1980", London: Thames and Hudson.

Frederick, C (1929) *Selling Mrs. Consumer*, New York: Business Bourse.

Gann, D, Barlow, J and Venables, T (1999) *Digital Futures: Making Homes Smarter*, Coventry: Chartered Institute of Housing.

Gibbs, WW (2000) "As We May Live", *Scientific American*, November, pp. 26–28.

Handford, R (2002) "Turn Up the Heating – From a Distance", *Financial Times*, London.

Hardyment, C (1988) *From Mangle to Microwave: The Mechanisation of Household Work*, Oxford: Polity Press.

Hindus, D (1999) "The Importance of Homes in Technology Research", *Co-operative Buildings Lecture Notes in Computer Science*, Vol. 1670, pp. 199–207.

Jedamzik, M (2001) "Smart House: A Usable Dialog System for the Control of Technical Systems by Gesture Recognition in Home Environments", http://Is7-www.cs.umi-dortmund.de/research/gesture/argus/intelligent-home.html

Junestrand, S and Tollmar, K (1999) "Video Mediated Communication for Domestic Environments: Architectural and Technological Design", in N Streiz, J Siegel, V Hartkopf and S Konomi (eds.), *Cooperative Buildings: Integrating Information, Organizations and Architecture, Proceedings of CoBuild'99*. LNCS 1670, pp. 176–89, Springer.

Kidd, CD, Abowd, GD, Atkeson, CG, Essa, IA, MacIntyre, B, Mynatt, E and Starner, TE (1999) "The Aware Home: A Living Laboratory for Ubiquitous Computing Research", in N Streiz, S Konomi and H-J Burkhardt (eds.), *Cooperative Buildings: Integrating Information, Organization and Architecture, Proceedings of CoBuild'98*. LNCS 1370, pp. 190–97, Springer.

Mateas, M, Salvador, T, Scholtz, J and Sorensen, D (1996) "Engineering Ethnography in the Home", *CHI 96 Electronic Proceedings*, http://www.acm.org/sigchi/chi96/proceedings/shortpap/Mateas/mm_txt.html

Meyer, S and Schulze, E (1996) "The Smart Home in the 1990s. Acceptance and Future Usage in Private Households in Europe", in *The Smart Home: Research Perspectives, The European Media Technology and Everyday Life Network (EMTEL)*, Working Paper No. 1, University of Sussex, Brighton.

Mozer, MC (1998) "The Neural Network House: An Environment that Adapts to its Inhabitants", in M Coen (ed.), *Proceedings of the American Association for Artificial Intelligence Spring Symposium*, pp. 100–14, Menlo Park, CA: AAAI Press.

Mynatt, ED, Rowan, J, Craighill, S and Jacobs, A (2001) "Digital Family Portraits: Providing Peace of Mind for Extended Family Members", *Proceedings of the 2001 ACM Conference on Human Factors in Computing Systems (CHI 2001)*, pp. 333–40.

Norman, D (1998) *The Invisible Computer*, Cambridge, MA: MIT Press.

O'Brien, J, Hughes, J, Ackerman, M and Hindus, M (1996) "Workshop on Extending CSCW into Domestic Environments", in *Proceedings of CSCW '96*, November, p. 1.

Sacks, H (1992) "Aspects of the Sequential Organization of Conversation", in G Jefferson (ed.), *Lectures on Conversation*, Vol. 1, Oxford: Basil Blackwell.

Scholtz, J, Mateas, M, Salvador, T, Scholtz, J and Sorensen, D (1996) "SIG on User Requirements Analysis for the Home", in *Proceedings of the CHI '96 Conference Companion*, p. 326.

Siio, I, Rowan, J and Mynatt, E (2002) "Peek-a-Drawer: Communication by Furniture", interactive poster, Human Factors in Computing Systems (CHI 2002).

Tang, P, Gann, D, and Curry, R (2000) *Telecare: New Ideas for Care and Support @ Home*, Bristol: Policy Press.

Tollmar, K and Junestrand, S (1998) "Workshop on Understanding Professional Work in Domestic Environments", in *Proceedings of CSCW '98*, November, p. 415.

Voida, A and Mynatt, ED (2002) "Grounding Design in Values", a position paper for the workshop on New Technologies for Families at the ACM Conference on Human Factors in Computing Systems (CHI 2002).

Wajcman, J (1991) *Feminism Confronts Technology*, Cambridge: Polity Press.

Films

Dream House (also released under the title "H.E.L.E.N.") (1998) Director: Graeme Campbell.

Demon Seed (1977) Director: Donald Cammell.

Households as Morally Ordered Communities: Explorations in the Dynamics of Domestic Life

3

John D. Strain

3.1 Preamble

Homes are not merely repositories of human action, nor can they be understood as the output of larger sociological phenomena, as is often argued in the mainstream sociological literature. Rather, households (or families or groups of people sharing common facilities – however they might be labelled) are in many ways communities in their own right: they possess a tradition, a moral order which frames and guides behaviour as well as the use of household facilities and technologies. This chapter explores, using ethnographic evidence drawn from a panel of households in the south-east of England, how households exhibit these patterns of distinctive moral ordering. It will explore the problems of understanding them analytically; and on this basis, point towards how this moral ordering needs to be accommodated when trying to design and implement digital technologies in and for home settings. A particular theme will be to explore, on the basis of evidence, how the adoption of digital technologies will, as a matter of course, generate various conflicts; but not necessarily of power and hierarchy, having to do much more often with such everyday matters as "appropriate behaviours" for different times and places. Here, issues to do with how the home is at once a place for hosting friends and for resting, for living out personal desires and for sharing with others who may have different desires, need to be managed and "worked out". Though these matters may seem from certain views prosaic, the conflicts that can result are often consequential. More importantly, they also have implications for what are the factors affecting the suitability of various digital technologies.

3.2 Theoretical Context

Giddens (1984: 119) uses the concept of "regionalisation" to refer to the zoning of time and space in routine social practices. The private house, he reminds us, is a locale serving many clusters of interactions in a typical day. Houses are regionalised into floors, halls and rooms which are zoned differently in time as well as space, night time serving as "the most fundamental zoning demarcation between the intensity of social life and its relaxation" despite the capabilities of artificial lighting. These particularities about how space and time is ordered in the household form the enabling constituents of understanding the household as what Silverstone et al. (1992) call the moral economy of the household. It entails recognising the household as a transactional system, part of the purpose of which is to sustain its autonomy and identity as an economic social and cultural unit. In their practices of managing incomes, purchasing goods and services and, more crucially perhaps, conducting those activities which individuals deem most constitutive of their identities and relationships with others, members of the household maintain what Giddens (1984) defines as their 'ontological security' – the sense of confidence that the world really is at it appears to be. And to these three processes of managing incomes, managing expenditure and living their lives, a fourth might be added: managing the interface between the legally expressed boundaries within which society requires activities to be conducted, and the necessary interpretations of these laws within each home – in short, shaping and being shaped by the various arms of governance; local, national and international.

The work of both Giddens and Silverstone et al. serves to remind us that homes are not merely repositories of human action nor can they be understood as the output of larger sociological phenomena, as is freqently argued in the mainstream sociological literature. Rather, households (families or groups of people in some form of relationship with each other and sharing common facilities) are in many respects communities in their own right. They possess a tradition, a moral order which frames and guides behaviour including the use of household facilities and technologies. In sketching out a model for understanding the relationship between private households and public worlds, Silverstone et al. (1992) faced the conundrum that household communication and information technologies are not only objects of use like dishwashers or video recorders, they are media. They conclude from this that information and communication technologies have the capacity to threaten the process of creating ontological security in that they speak of events and objects which are separate, in both time and place, from events and meanings which constitute the moral order of the home. The telephone facilitates conversations which constitute social networks beyond the home; but such conversations can also challenge meanings derived from relationships within the home. The television and the computer transform the boundaries of what is deemed important in the home and offer the possibilities of change.

This extension of the boundaries of meaning for the household is no doubt a significant challenge to them. But it would be misleading to suggest that these are the only challenges to the moral economy of the home. With or without new digital communication media and technologies, people come in and out of households, for work, for learning and for play, all of which create networks of meaning which might constitute their own challenges. The conflicts that can ensue might constitute challenges to power and hierarchy; but not necessarily so. Sometimes these conflicts can be of more prosaic issues involved in the adoption of digital technologies, conflicts over mundane matters as "appropriate behaviours" for different times and places. Here, issues to do with how the home is at once a place for hosting friends and for resting, for living out personal desires and for sharing with others who may have different desires, need to be managed and "worked out".

Discussion of the boundaries of the household raises the question of the ontological status of households. One of the key critiques of the "systems approaches" to analysing social entities, whether of the more general kind of structural functionalism of Parsons (1951) or the more discrete variants of socio-technical systems analysis is that they are inclined towards "normalising" harmony and common purpose and treating conflict as pathologically dysfunctional (see Silverman, 1970, for a critique of "systems" approaches to organisations). There is perhaps a similar danger of reifying the moral economy or moral order of households to being something which can be defined without reference to the persistent and chronic negotiation of the moral order which takes place in households, with or without the incursion of digital technologies. The aim of this chapter is to explore, using ethnographic evidence drawn from a panel of households in the south-east of England, how households can exhibit patterns of distinctive moral ordering in household practices of managing their incomes through domestic banking systems and gathering their household resources through shopping.

The argument I wish to develop in this chapter requires a distinction to be drawn between two different, but related levels at which mundane activities such as banking and shopping need to be understood. At one level they need to be understood anthropologically as components of what constitutes people being in relationship to each other: responsibility, commitment, need, affection and love, all playing some part. These attributes of human beings in relationships have an inextricably ethical dimension to them. At another level, they need to be understood as transactional processes in which access to information plays a key part. How this information is sought and managed will reflect ethical choices by members of households which need to be recognised if our understanding of household practices is to serve the design and implementation of digital technologies in the home. What I wish to argue is that the frameworks of ethical choice or the moral order of the home is susceptible to change of a particular character. The moral order of the home is neither in a state of continuous flux in which every transaction between householders and the environment represents a worrisome new challenge to the moral and ethical order of the household; nor is this moral order impervious to the affordances of technological devices offered to households as media for conducting their banking and shopping. What I do want to suggest is that paradigms of the moral order in households reflect both stability and significant change as technological devices are appropriated. The affordances of new devices are frequently accommodated within a paradigm of the moral order; or they may be rejected on account of an incompatibility within that moral order. Alternatively, the moral order may undergo significant change whereby devices and their affordances come to be accommodated within a new paradigm.

The thesis has considerable significance for the design of devices for households. Devices are frequently designed on the basis of particular models of how households might or might not use the devices, models generated by marketing data and by understandings of the preferences of potential customers. But the affordances of the devices once designed may have significant impact on the moral order of the home to the effect that the devices come to be used, or rejected, in ways quite contrary to the expectations of designers. There is a case therefore for ethnographic pilot studies of new devices within households that explore the manner in which these shifts in the paradigms of the moral order can, and do take place.

3.2.1 Morality and the Moral Order

There is a temptation for philosophers, if not sociologists, to succumb to the notion that there is some timeless and determinate category of moral concepts that is independent of how people behave in any place or time. As Macintyre (1998) pointed out, "moral concepts are embodied in and are partially constitutive of forms of social life". This is not to suggest that there can be no criteria by which the practices of any society or group within a society might be evaluated. But it does entail that if we are to understand how people make sense of digital technologies in households, some purchase is required on how people within households construe what is right, appropriate and good in respect of themselves and their households. Again, it does not imply that people's values and principles are unchanging, and the research reported here on the use of Internet shopping and banking revealed the strength of feeling that was attached to how information technology, in its broadest sense, could impinge upon deep-seated notions about what was right and wrong in respect of household practices. Unsurprisingly, perhaps, the sense of how technologies impinged upon household life varied considerably with the character of the household. People in households with children under eighteen displayed more eagerness to define a moral order for their households than those without them. But in no case was it impossible for some definition of this moral order to be made.

This study conducted by the Digital World Research Centre into patterns of information use in households enabled the construction of a matrix of constraints and opportunities for what Norman (1988) called the affordances of digital technologies. It served to ground some of the more unrealistic aspirations of those who dream that the "Martini"[2] solution to providing information in the constraints set by how information is actually used within households. By making the household the focus of the study, it served to identify constraints that emerge only by considering relationships within households, rather than on individuals. But it also allowed the extent of how the moral order of the home is persistently renegotiated by its members to be revealed.

The study in question comprised an extensive series of interviews over a twelve-month period with members of twelve households. They comprised three groups: the first included retired people; the second group comprised households of adults below retirement age and with children and teenagers; the third group comprised adults at work without children. The study identified the way in which information is procured, manipulated, transferred and debated within households in relation to shopping, banking, governance and "infotainment". The extended period over which the interviews were conducted allowed considerable knowledge to be gleaned about how practices within each household related to the sense of moral order in the home; and more saliently how agreement over this moral order was chronically renegotiated and how the affordances of digitally mediated information sources affected the quest for agreement over this order.

In extremis, of course, if there is no agreement, there is no order. But here again Macintyre's treatment of socially embedded moral orders is illuminating. Macintyre's project began with a history of ethics which sees ethical practices embedded in, and not disassociable from, history. The evaluative concepts, maxims, arguments and judgements that comprise a morality are nowhere to be found except as embodied in the lives of a particular people. "Morality which is no particular society's morality is to be found nowhere" (1981: 266). Central within Macintyre's project is the work of Aristotle for whom virtues are central to ethics as the habitual dispositions which both facilitate and contribute to human goodness. The practice of a virtue is not simply a means to an end, it is itself part of what counts as contributing to what is good. For moral truth can neither be isolated from claims about the purposes of human life, nor from the social contexts in which these are pursued.

Macintyre, following Aristotle, focuses upon the social practices which virtues serve, and the meaning of a whole life within which these social practices can be shaped and harmonised. This shaping process depends

---

[2] The so-called "Martini" solution refers to the availability of information any time and at any place.

upon a particular history or tradition within which priorities can be set. Human beings and the narratives they forge are governed by a narrative unity. Macintyre provides an account, therefore, of how morality develops in particular societies and in their particular histories in which relation to what each society seeks to achieve and the virtues and practices which promote this.

The significance of Macintyre's account (developed further in Macintyre, 1998) is that it is capable of showing not only how grand changes in the moral landscape of history take place; but also perhaps of how more prosaic changes in the paradigms of moral order take place in homes. Macintyre's aim was to show how the holders of a particular moral tradition can experience irresolvable conflicts within their own traditions. In such circumstances of crisis, alternative constructions of what is crucial within the tradition will arise. These new constructions of what is important may use the language of different, rival traditions, thereby allowing different traditions to serve in the way second languages work. Second languages, with their different vocabularies, may enable things to be said in ways which are difficult or awkward in one's first language. Gradually, exposure to new practices and opportunities provides a language within which older traditions can be re-expressed and valued. As Fergusson (1998) explains, it is possible for "a tradition's ability to use language to accommodate the insights of another [tradition] while also resolving new problems which are incapable of resolution in the rival account." Not only can we come to learn a second language, we can sometimes find that certain things can be said in the second language that cannot be said in the first. For language, substitute moral discourse, and there is possibility of changing a moral tradition by means of the insights provided by another.

Perhaps a simpler account of this process of negotiation and resolution is provided by Holloway (2001) in his account of "ethical jazz". In an attempt to steer a course between a moral fundamentalism ("defending a tradition in a traditional way" as Giddens put it) and a relativism that rejects any authority within a tradition, Holloway suggests that "ethical jazz" in which a certain competence to play music is treasured by its practitioners, but what is played in any event is inherently improvised. What Holloway recognises is that paradigm shifts, as Kuhn (1962) described them, can take place in the moral sphere.

The focus of Macintyre's analysis of moral frameworks lies in grander scale historical traditions than the prosaic practices and moral orders of particular households. His concern is with communities of traditions of thought, such as those associated with liberalism, utilitarianism and other inheritances of what he considers to be the false and unworkable notion of reason in relation to the ethics which the Enlightenment project bequeathed. Indeed Macintyre (1981) suggests that contemporary morality in practice is in such a state of chaos, that we cannot even recognise the seriousness of it, let alone provide resources to deal with it. I wish to suggest however, that the possibilities he describes of exchanging understanding across rival traditions of thought is evident at a far more prosaic level than in the grander traditions of ideas he describes. In the practices of banking and shopping in the households of south-east England can be observed phenomena of change which are closely parallel to the possibilities of change that Macintyre describes in the history of philosophy.

Before turning to these examples, it might be well to briefly defend the Macintyre thesis from the charge of relativism in respect of morality – that statements of how people ought to act can have no truth value in themselves, but be truth bearing only as representations of how people regard things. It is true that Macintyre rejects that tradition of ethical reasoning, inherited from the European Enlightenment, which sought to define a tradition independent yardstick by which to assess rival constructions of people's moral traditions – a meta-ethical theory of people's ethics as it were. This might suggest that people's ethical traditions are incommensurable and incapable therefore of bearing judgement against them. But Macintyre was at pains to suggest that the absence of a 'tradition independent' yardstick does not rule out an indirect commensurability of rival accounts of the ethical order. Although traditions are, 'sub specie aeternitatis' incommensurable, translation is possible between one tradition and another, thereby facilitating their mutual critique.

3.3 Analysing the Practices – Continuity and Change

An example of how changes in language within which moral discourse is conducted can take place in the mundane practices of the home, as well as the grander scale that Macintyre describes, can be found in considering household practices of banking. But before doing so, some consideration of the case for focusing on the dynamics of change within households is appropriate in the context of existing literature on shopping and banking.

Much of the effort in understanding user needs for information and communications technology from an ethnographic standpoint has taken place in work rather than domestic contexts (Bannon and Schmidt, 1991). Frohlich et al. (1997) drew attention to the lack of research into domestic requirements for new broadband communications technology in a call for effort to be directed to domestic contexts in what they describe as "computer supported social inter-action". Mateas et al. (1996) provided a powerful example of the contribution to be made by ethnographic study of the home. Their study of ten families revealed the complexities of space, time and communication in home life and demonstrated the inappropriateness of many of the characteristics of the personal computer for the home. Whereas people cluster in what the authors called the "command and control" regions of the home – the kitchen and family room – the computer is typically found in a personal work space, designed for use in a single space with clear demarcation between work and non-work between being "booted up and working" or switched off.

Approaches that focus on domestic contexts might be seen as encompassing three types of approach. The first has focused on domestic activities themselves rather than their digital mediation. Within this approach there has been a consideration of traditional shopping journeys, assisted by in-store digital technology. Hopping (2000) has examined the technologies available to retailers and provided a summary of possible changes in retailing in the next five to eight years. His approach might be considered a technocentric one in so far as it treats changes in retailing as a consequence of technological change. But he recognises that these changes are constrained by choices made by both customers and retail organisations, many of them in pursuit of cost reductions by such techniques as bar code scanning and transitions from cash to electronic funds transfer.

While in-store technologies reflect considerable capability, even greater technological advance is evident in developments in the Internet. Sixteen million people make use of the Internet in the UK as of 2001, according to a survey by the UK Consumers' Association: 36 per cent of the UK population go online – an increase from 27 per cent a year earlier. With bandwidths 1000 times faster than today enabling the alleged convergence of video, audio and data transactions, major increases in the use of the Internet for banking and shopping is technologically feasible. But as Hopping points out, this is unlikely to generate an increase in Internet shopping beyond about 20 per cent of the population buying 20 per cent of their goods on the web. Although almost 8 million people have now shopped online, and the variety of products purchased has increased, for 87 per cent of those who have used Internet for shopping, the range of goods is limited to books, CDs and software. It is possible of course that far more people used the Internet for investigation or enquiries leading to a purchase, without concluding the transaction on the Internet.

The second approach attempts to explain the dynamics of the use of Internet shopping or banking itself, and includes Rowley (2000) who provides a research agenda by which we might understand better the way in which people use the Internet for shopping. She makes use of a model of consumer buying proposed by Engel et al. (1990) and discussed in marketing texts such as Brassington and Pettitt (1997). This model is a linear series of rational steps that makes explicit the processes of recognising a problem by a consumer, identifying what sort of purchase might solve the problem, identifying where and how such a purchase could be made, obtaining information to make a decision about a purchase, making the decision and evaluating it afterwards. Rowley recognises that such a model simplifies the myriad influences on the decision-making process and the multiplicity of types of purchasing which include impulse buying, fashion following and routine response behaviour. One key problem with such rationalising models is that it is not clear that the full range of these influences can be accommodated within such a rational model without doing violence to the nature of the behaviours themselves. While it may be the case that people will wish to provide "rational" explanations of their behaviour, post hoc, it does not follow that such explanations figure at all in the purchasing decisions.

One method of exploring what people strive for is to examine the rationale people use for adopting particular practices. The theory of planned behaviour (Benham and Raymond, 1996) is one approach to examining the rationale for adopting practices embracing attitudes, subjective norms and perceived behavioural control. Within this approach, an examination of the adoption of virtual banking in Hong Kong has been made by Liao et al. (1999) making use of the theory of planned behaviour. Virtual banking was defined widely and included the regular use of ATMs, phone banking and home banking making use of PCs and/or Internet. She found that a number of hypotheses drawn from relationships predicted by the theory of planned behaviour tested positively.

The risk to be faced, however, by strongly prescriptive models of choice is that they run counter to the evidence of much social enquiry. People do plan but the evidence is they plan much less coherently than strongly rationalist models suggest. At the heart of the sociological debate about choice and intention lies the question of the degree of planfulness in people's lives. Anderson et al. (1994) provide the first major study using survey methods of household plans and strategies. The evidence suggests that there are different sets of strategies that apply to the world of work and the world of home. But in both worlds, the concept of strategy must not be "reified". Plans and strategies are constructs people use to make sense of the world. People do define goals and they have more or less rational plans for achieving them. But goals and plans may cohere loosely at a high level and serve to help people cope with the circumstances as they present themselves to people at home, rather than serve as dynamos of action.

This brings us to the third approach, routed in the tradition of ethnographic study. O'Hara and Perry (2001) point out that e-commerce transactions are currently concentrated in relatively standardised transactions such as books and CDs and for what they describe as "intention driven" shopping. They suggest that online shopping will plateau at about 20 per cent of retail sales partly because of the need to sense and experience consumer objects more completely than is possible online; and partly because the broad movement of digitisation goes far beyond the confines of the Internet-enabled PC. It looks towards a mobile and ubiquitous revolution in which people move, sense and interact with their environment, with varying impulses to act and to consume. Where impulses cannot be acted on there exists what has been called an *intention-action* gap, created by a number to types of discontinuity: physical, where the person is not in the right place to act; informational, where there is insufficient information to act; and awareness discontinuity, where the opportunity to fulfil the impulse is outside the focus of attention. Recent mobile technologies provide opportunities to bridge these various discontinuities. O'Hara and Perry argue that many of these new mobile devices tend to mimic desk-bound transactional devices and fail to address the full scope of the shopping experience which better designed devices might otherwise support.

This points to the need for an understanding of the shopping experience itself. This has been met in part by recent ethnographic work in shopping such as Miller (1998) and in part by the attempts of marketing strategists such as Underhill (2000) to enable retailers to understand better the psychology of their customers. While it is commonly assumed that shopping is primarily concerned with individualistic materialism, Miller rejects this assumption, and analyses shopping by analogy with anthropological studies of sacrificial ritual. The act of purchasing goods, he suggests, is almost always linked to other social relationships in which issues of love and care for others have a central place. While this particular variant of ethnographic study is wide ranging in its coverage of the shopping experience, it has little explicit to say about the implications for the design of computing technologies. O'Hara and Perry's work was intended to fill this gap. Their study of shopping required people to record the occasions on which they both experienced the impulse to buy, and, for whatever reason, had to postpone the purchase. By exploring the reasons for this deferral, the authors suggest a range of possible devices which might meet the affordances sought for by consumers in their shopping experience, but for whatever reason, denied. They suggest too that rather than characterising consumer behaviour in terms of models of rational processes, they might conform better to what might be called "consumption narratives", from an initial impulse to a final transaction, a narrative which includes an initial awareness building and the articulation of "wish lists" prior to consumption.

Ethnographic investigation such as the above indicates the limitations of home shopping on the Internet. The UCLA Internet Report (2000: 46), *Surveying the Digital Future*, indicates that a very small group of purchasers is responsible for a large proportion of the purchasing. The UK Family Expenditure Survey of 2001 indicates that of those in the UK making purchases via the Internet, purchases were frequently limited to books, CDs magazines, travel tickets and accommodation. This is not to say of course that Internet shopping does not have an important future. The same report also suggests that more than half of Internet purchasers (54 per cent) believe that they will eventually make many more purchases online.

As Jones and Biasiotto (1999) point out, some advocates suggest that e-retailing will account for 55 per cent of all purchases by 2015. Others consider the hype to outweigh by far the real potential. These authors themselves take a different approach to either of the three discussed above, to exploring Internet shopping. Instead of exploring the perspective of consumers, they assess the current use of the Internet as a retail option for location-based retailers. They provide a wider ranging review of the experiences of 39 cyber-retailers with Internet shopping. They conclude that many of the concerns associated with the technology (security, speed of access, branding, product availability and distribution) are being addressed and consumer acceptance will accelerate rapidly over the next five years. On the other hand they argue that it may take a generation for e-retailing to achieve mainstream status. It will not happen until "the typical middle aged consumer feels comfortable with the new technology".

A more detailed scrutiny of the emotions involved in shopping emerges in the study by Omar and Kent (2001) of the influences on impulsive shopping. Their acute analysis of the somewhat unique features and practices of airport shopping point to a range of factors relevant to the affordances of shopping more generally. "Impulsive shopping" is a consumer trait by which shoppers respond to sudden unexpected buying ideas. Whether or not these ideas result in purchases depends on intervening factors. These factors include consumers' own subjective evaluations of acting on their impulses and on normative evaluations. Subjective evaluations range from seeing them negatively as irrational, wasteful and shortsighted, to seeing them positively as pleasurable, responsive to opportunities and (in cases of gift purchases) generous and thoughtful.

The analysis of the emotions associated with shopping or banking may have an important role illuminating the ethnographic context. But emotions can be fickle things and they suggest a focus on the processes of change within people as they come to see opportunities within different perspectives. Issues of change and new perspectives were an important dimension of my own study.

The concern underpinning the chapters in this book and not just my own, is one which emphasises the real needs of people in particular contexts, rather than with the concerns of, for example, sociologists or indeed any other discipline predisposed to explaining phenomena in terms of concepts remote from those articulated in daily life – concepts such as modernity or alienation. But in addressing the real needs of real people, it may be important not to lose sight of the real beliefs people hold about how their lives should be conducted, beliefs that may be as important to them as the practices. Our investigation of shopping and banking practices was informed by a concern not to lose sight of what people believed about what they were doing without losing focus on what they actually did.

3.4 Changing the Paradigm of Banking

When the language of banking is changed to reflect changes in practices from activities such as "writing to the bank manager" or "visiting the bank" to language about "going online" and "downloading a balance", more than mere words are at stake. As I found during my observations of the shift from bricks and mortar banks to e-banking, opportunities are created for people within households to readjust roles within the home which change the moral order. In three of the twelve households, banking was characterised as that of trusting the professionals, that is, the major banking institutions. This was evident in one household where the partners had quite different conceptions of banking and where the process of renegotiation was apparent. The husband, who had no interest in having a home computer, had no confidence that good service could be obtained outside a small branch of a bank with whom a longstanding relationship could be maintained. The female partner was eager to have a home computer and eager to explore the scope for saving time by easier access to accounts. But the debate between them was about much more than the utility of saved time. It was a debate between what gave the man confidence that his affairs were orderly with respect to the world and what was offered to a woman who had no need of such external support to the validity of her life, dominated as it was by her role as carer and mother to her disabled son and his brother. What was at stake was a shift of paradigm about the moral order, a shift which was facilitated by the apparent affordances of digitally mediated information.

This shift of paradigm was more apparent in the case of another household with a teenage child, a household where a former greengrocer and publican was the father of the household with very traditional understanding of his role as breadwinner and guardian of the household's security. The woman described how the role of maintaining bank accounts had long been held by the husband, although she had been routinely checking and reconciling balances with outstanding cheques. The couple's approach to banking for the first twenty years of their marriage was categorised in the study as one of "reliance on trusted professionals". They had never changed their bank in twenty years of marriage, regardless of occasional errors by the bank and frustrations with the service provided. Their rationale for staying with the bank had been that relationships with banks are "for the long term". To change banks represented a loss of trust in an institution that would have radically threatened their trust in a whole network of financial providers which extended beyond banks. They had introduced their teenage daughter to the same bank. Investment of their faith in their bank was an expression and embodiment of a core feature of a moral order in society which helped define the integrity of their household. It reflected what might be called a paradigm of the moral order linking the household to the banking system which their household management required.

And it was the father's role to safeguard this practice. Their gradual introduction to Internet banking was conceived as entirely within this paradigm. It was simply a way of accessing balances more quickly. Gradually, however, the woman came to see a wider range of financial services and products available on the Internet, which she initially avoided on account of her lack of trust in the Internet. But, as a result of using the Internet regularly, she gained trust in it to the point where the investment of faith in the banking system was replaced by an investment of trust in the Internet itself. A key moment came when, as she explained, she shifted funds from a deposit account with her bank to a new bank operating on the Internet. Over time, she came to see that her key relationship was not so much with a bank but with a medium, the Internet, which gave her access to a range of banking services and tools to give her ready access to the family's balances. And from that point she changed her bank and created several new banking relationships whereas the family had remained with a sole banker for twenty years.

Part of this process of transformation was a process of slow negotiation within the family. Prior to her taking on the banking role, the husband had been seen by both partners as the guardian of the family finances, maintaining a long-term trusted relationship with a bank, with his wife occupying a supporting role in checking balances. This role was consistent with his role as a father and guardian within the family across a wide range of activities which, in an Aristotelian sense could be described as his manifestation of virtuous living in the moral order of the home. But without the skills necessary to operate the Internet, there was a process of negotiation whereby she came to occupy the role of financial manager, with the husband supporting her and giving her confidence in the process.

This particular case was interesting because there was a paradigm shift, obtained through negotiation, in how the banking function was regarded in the family. But a comparable sense of banking being part of a moral order was evident in another example of a household with two parents and two children. For this family, the woman had always been the financial manager, supported in this role by the husband who deferred regularly to her judgement. They had always fiercely resisted credit except for their mortgage and dealt wherever possible in cash. Interestingly both parents were very technologically oriented and made use of the Internet regularly for information quests. Their resistance to Internet banking was rooted partly in fears over security but partly in their eagerness to limit the role of banking in their lives. For this family, when contacts with banks were needed, personal contact with the bank and negotiations conducted with clarity, honesty and integrity in efficiently run offices were of far more importance than digitally mediated transactions. A key part of their rationale for this lay in the example they wanted to set for their children. Life for this couple was not a rehearsal for themselves, but a key goal of their life was to live it in way that would create role models and patterns of good living for their two children. This was also a key factor in how they did their shopping, about which, more later. The point to be laboured is that in three of the twelve households, banking could be categorised as the maintenance of good relationships with trusted professionals. But the arrival of digital technology in all of these homes created opportunities for this aspect of the moral order of the home to be renegotiated within the household into one where the bank occupied the role of "information servant" to financial management within the home.

Nine of the twelve households demonstrated a different approach to banking from the outset of the study, one that was categorised as maintaining a "master-servant" relationship in which the bank was regarded as a servant supplying the information needs of the household's banking. In these households, the introduction of Internet banking proceeded more smoothly, but an interesting paradox emerged in relation to the services offered by the banks. Many of the services were offered as part of a comprehensive package which might secure stronger customer loyalty from customers amenable to Internet banking. But many of those who turned to Internet banking most easily were those who had a strongly "instrumental approach" to banking. They wanted particular services, accounts and deposit mechanisms as part of a financial strategy which was their own, rather than the banks. Their medium for investigating what was on offer was not the bank but the Internet itself which offered services and tools from a wide range of suppliers. What was seen by the banks as a strategy for securing customer loyalty via the Internet proved in its outcome to threaten this customer loyalty. It is a demonstration of the need to take account when designing services, of how the services will be regarded by consumers in their vision of domestic ordering, although in many of the cases where Internet banking was used, the adoption of a master-servant relationship to banking had come to be shared by both adults in the family. This sharing of a common categorisation could mask other differences. Among one of the childless couples, the male partner was a veteran Internet banker, valuing the scope to have instant access to balances and confirmation of transactions. His female partner took a similar "master-servant" relationship to banks, but with a greater intensity. She was a professionally qualified accountant, managing a family garage business. Her mode of accounting for domestic money was to hold money only in deposit accounts and to use only cash for shopping. She saw this as a highly effective way of managing the household's money and maximising spending power by being able to secure price reductions. It was a radically different approach to her partner who wanted to able to transfer money between different accounts and make use of credit cards. The differences between them in their approach to money was managed by establishing quite different responsibilities for who would pay for particular services. The female partner kept the home stocked with food. The male partner paid "the bills", that is, for utilities and mortgage payments. This was also a young household, of a childless couple who had been together less than ten years who expected that their modes of managing money might change in the future. Managing differences in responsibilities for shopping by this particular couple was more easily managed than managing differences in approach to financial products, such as buying into investment trusts. For many couples in the panel purchasing investment products such as these over the Internet was seen as unproblematic. But for this couple, there was acute difference between the man and the woman. For the man this was a straightforward and effective use for the Internet. For the woman, for whom face-to-face conversations with a trusted advisor explaining tax implications in "plain English" is part of her schema of how things are done with integrity, it was deeply problematic. The result was that no purchases of these products online were made. This was less a matter of the locus of power in the relationship, but a manifestation of how in financial affairs, the fears of one partner for a project will often outweigh the enthusiasm of the other. In joint financial projects, the fears of either party can threaten the joint project. In purchasing material objects, by contrast, it is easier to manage differences by apportioning purchasing responsibilities for items. It is in these ordinary, apparently prosaic activities, transferring funds, paying credit card bills, and saving for the future, that the variety of forms constitutive of family structures manifest themselves. It is in regard to these differences that the shaping of new technologies occur.

3.4.1 Shifting the Shopping Paradigm

Different approaches to banking within households required the management of important differences among attitudes of people within households. These were even more apparent in respect of shopping among the panel. Even more strongly, however, the patterns of shopping and provisioning reflected deeply held convictions and concerns bound up with the nature of household life and the human relationships within them.

The presence of a computer in the household was not a criterion for selection to the panel. Yet, there was a personal computer in eleven of the twelve households, suggesting a high level of interest in digitally mediated services. It was somewhat surprising, therefore, that so few of the households made any significant level of purchasing over the Internet.

The accounts given of how and why people went shopping provides some clues in relation to people's sense of moral order. The first theme that emerged was the persistence of negotiation over shopping. Of the eight households with children and teenagers, this theme was strong in six of them but it manifested itself in various modes. In a single parent household, the local and very nearby nationally branded store served as the household's larder. Very little was stored in the house and either the mother or one of the two boys visited the store almost every day to choose what they wanted for their meals. The small size of the totals purchased each day meant that negotiations could take place before each trip in a relatively small amount of time. The negotiations took the form of compromises over what was eaten. Hamburgers could be bought and eaten one day provided that a "healthier" meal was eaten the next. But frequently the agreement was impossible to maintain, perhaps because a certain item was not available, perhaps because circumstances changed in the time between the deal struck at breakfast time and the shopping trip after school. So a further deal would be made the following day which took account of the breach.

What is of particular interest here is twofold. First, the basis of this store serving as an arena of family negotiation was the contingent proximity of the particular store with the range of products it offered. Had it been much further away, it could not have served in the way it did. Secondly, the pattern of somewhat conflicted negotiations in the family were part of a wider pattern of emerging conflicts between two very different boys and their single parent mother. Some years earlier, when both boys were under ten years of age, there was quite a different moral order in the home, in which the mother shopped less frequently and made more decisions by herself, consulting rather than negotiating with the boys. After all, the boys were growing up. During the course of the study, the role the boys took in shopping increased as they took a greater part in selecting items for purchase. What was apparent, therefore, was that a particular type and location of store provided an arena which facilitated a small but nonetheless significant transition in the moral order from one of limited parent lead consultation to one of extended negotiation, one in which quite different products were bought at quite different frequencies. Here the boys were increasingly participating in "grown up affairs" albeit in a small way.

The above example of transition in the moral order involved no new information technology or device. In another household there was a difference in opinion over the potential role of the Internet. The male partner, although technologically capable, resisted Internet shopping for groceries. The female partner was eager to make use of it, but felt prevented by her partner's lack of interest. There was also persistent disagreement between their approaches to shopping. The male partner preferred to do the shopping provided he could do it speedily, system- atically and with a degree of sensitivity to his personal favourites. He regularly shopped for the household in a nearby supermarket; but he had less regard for what the female partner considered as "healthy eating" for her two children. The couple felt that their pattern was, although disputed, an acceptable compromise. The female partner was too busy to spend enough time in the store and she valued the contribution of effort her partner made.

In discussion about how things might be, however, the female partner was adamant that were she able to make use of Internet shopping, she would take far more control over what was bought, in terms of quality and ingredients. She made it clear that her beliefs about what was important in buying much healthier food for the two boys would become stronger. She didn't feel that she was betraying her principles in not making a stronger stand over her partner in not paying enough attention to the ingredients in food purchased. She simply recognised that putting her beliefs about healthy food into practice would simply make more sense in her life, if she had greater control through the Internet. It was not certain to her that using existing Internet grocery shopping would necessarily give her sufficient knowledge of food ingredients. Internet shopping facilities simply didn't provide that level of product detail. But if they did, she would have valued it.

It would appear, therefore, that if an Internet shopping facility was designed for them on the basis of their existing practices, it would be designed on the basis of the male partner's practice. The female partner had no practises, only a set of beliefs. So a design based on practice would have addressed his need to navigate speedily between categories of shopping. It would thereby fail to address the potential for a mode of shopping which addressed the female partner's beliefs about the importance of ingredients. A design based on her beliefs would focus not on navigability but on depth of information on product quality. Thus, only by understanding the beliefs within the moral order can a design be truly reflective of household needs.

One might conclude from the evidence of disagreement and negotiation over shopping that shopping was generally disliked. Yet this was far from the case. If one reason for Internet shopping take up was lack of time for shopping, one would expect the take up to be greatest in routine grocery shopping. In fact, Internet shopping for provisions had very little take up among the households. And although all families could recall elements of discomfort with shopping, associated sometimes with lack of time and sometimes of the shops being too busy, the response of the shoppers was to find more convenient times for shopping, either during the weekdays, evenings or on Sundays.

In one household, going shopping was regularly done by the female partner with one or two of her two children. Her involving the children in the shopping had an educational role. In part it was a way of teaching children the value of things. Understanding what different things cost was part of bringing them up with an awareness of the relative cost of items. In part it had a more explicitly moral significance. There was "a way of behaving in shops". Not eating things before they had been paid for, taking turns in the queue, striking a balance between touching and feeling the quality of greengrocery without damaging the goods. All these formed part of the process of moral education that the woman felt had an applicability beyond the activity of shopping. But as well as this explicitly moral role, bringing the children with her was also for her the safest way of reminding herself of her children's needs in relation to what they would eat and what they would wear.

It might follow that this family were resistant to Internet shopping, but in fact both adults were deeply interested in technology. They were qualified amateur radio enthusiasts and the male partner was capable of repairing television sets and computers. When the female partner was about to enter hospital for an operation, she investigated the prospect of using the Internet for grocery shopping. Little headway was made in using the Internet on account of lack of choice available, the impossibility of testing the quality of fresh food and fears over credit card security. But they have a recognition that in time, they will use the Internet technology for shopping. They recognise that other moral issues, particularly over protecting their children from inappropriate material will become more important than using shops as templates of moral education for their children. Again, they envisage the possibility of a radical shift in what they see as the moral priorities for their household. Designers who might see their current practice as evidence of the inappropriateness of Internet grocery shopping might fail to recognise the possibility of a significant shift in their conception of the moral order that Internet shopping could provide.

A further example in the shift of the moral order is evident in the case of a retired couple with a twenty-year-old son studying information technology at degree level. For this couple, shopping had for over thirty years been a well-established practice in which the female partner visited a supermarket with a list compiled by the household. The male partner accompanied her only to help transport the goods in the car. The female partner did "top up" shopping, twice or three times a week, partly to support the local shops, and partly to greet and converse briefly with people she was confident of meeting in these brief perambulations. This same couple held a "master-servant" relationship with the bank of which they had been customers all their adult lives. The manner of shopping, particularly the "top up" shopping was a key part of their identities and their sense of the moral order for their home. Both partners came from large families who lived, and had lived for several generations, in the same area. They had very strong family ties had an equally strong sense of living their lives in a manner that was consistent with their parents' lives. Traditions, including church going and public service to the neighbourhood in the scouting movement were deeply precious to them. And this extended to the manner in which they behaved in patronising the local shops. The way they shopped was part of the way they lived, a matter which was part of a deeply held set of values.

But they were not without technological interests. They had used amateur "walkie-talkie" sets in scouting adventures, both in the field and to maintain contact between cars in convoy, long before mobile phones arrived. They were among the earliest users of mobile phones, and the male partner had worked for the nationalised telecommunications utility well before its privatisation. Their only son, who still lived at home during his degree and had a deep interest in information technology, multimedia and the Internet, was an important source of information to them both about the possibilities of using the Internet – for shopping. The male partner had long used a mail order catalogue, with a firm to which he had been introduced by his mother. The female partner had a passion for finding out information concerning her hobbies and interests. Both, but particularly the mother, were eager students of their son who was teaching them how to use the Internet. First, enquiries about "white goods" and prices were made on the Internet. Then some books had been ordered. Their use of the Internet made them aware of issues in relation to protecting children, issues they care about deeply. Gradually, it became clear to them that the Internet was a technology of the future, Controlling it to the benefit and well-being of people, making savings in their purchases to eke their pensions were all gradually emerging as more important issues than continuing to visit the supermarket and local shops.

Nothing in the above analysis is intended to suggest any sort of "technological imperative" towards using new technologies and devices. Many cases were identified where households could envisage no alternative to practices of shopping which required no use of new technology. For some couples, their preference for visiting stores rather than using the Internet was more concerned with shared pleasure. They enjoyed walking round the aisles together, debating and negotiating what they might eat together, what they would buy and how much they could save. One woman regretted the fact that her partner preferred to visit the supermarket alone and she took particular pleasure from going shopping in the town centre with him for non-perishable items.

Shopping was important to one woman because it provided space and time to allow commitment. The woman needed to shop alone for clothes because she found it difficult to come to a commitment about what to buy. Money pressed, but time rich, spending money on clothes represented a considerable commitment. How will it look? Will if fit? Will I wear it? were all questions she quietly worry about. So she would frequently wander between shops in a particular town and allow the travelling time between shops to act as a natural boundary of time in which to make a decision. Internet shopping was too open ended for her. Supposing, she was asked, the website provided its own boundaries of time to decide? An artificial time stop could remind the woman that time is up, that it's time to decide. But this wouldn't satisfy her. Who says it's time up? Who says it takes twenty minutes to get from one shop to another? Whatever the answer is, for this woman, she didn't feel it was something arbitrary. She wanted to make the decision when she felt she had to, bounded by constraints that appeared "natural" to her, rather than "electronically" artificial.

3.5 Conclusion

There are no grounds provided in the evidence I have presented for technological determinism. But what, I believe, there are grounds for is an understanding of how what appears to lie at the cornerstone of the moral order of household life can change significantly when perhaps gradually, perhaps suddenly, the import of all that is afforded by new technologies, makes what was once vital, seem no longer important. In the case of banking, there is evidence of Internet banking being seen as irrelevant in a household in which banking was "properly" conducted by the man in the house. An analysis of the affordances of Internet banking would initially have been expressed in terms of their significance to the husband who valued a personal relationship with his bank. The affordances of Internet banking would have been minimal. But to base the design of Internet banking on such affordances would be misleading. Such a design would have failed to register the scope for changing paradigms in banking. What at first was seen as a quite minor change in banking practice, the wife gaining Internet access to bank balances, were capable of being accommodated within an older paradigm in which the main decisions in banking lay with the husband. So accommodated, such affordances had little bearing on any decision to use the Internet for banking in a major way, or to change banks. But an accumulation of affordances gradually allowed a major shift in the paradigm in which the Internet was used to harness a range of different banks for different purposes with the woman in control of them. Any analysis of affordances after this shift would have quite different significance for the design of Internet banking services for the household.

A further difficulty in analysing affordances is evident where Internet banking encourages a shift from a relationship with a bank as a "trusted professional" to one in which the bank, or even, the Internet itself, is seen as a source of banking information and data. Several banking websites might well have been designed on the basis of deepening customer loyalty inherent in the bank's being regarded as a "trusted professional". What the website afforded to customers in its range of services from cradle to grave was initially accommodated within this paradigm and seen as a mark of the bank's professionalism. But the consequence of using the Internet for banking encouraged a view of the bank as a more limited provider of data which was equally accessible through other banks. A shift in the paradigm of how the bank was regarded meant that the affordances of a banking website were no longer relevant in the same way. It led to a loss of the very customer loyalty that the bank had sought to engender through its website.

Similarly in the shopping domain, Internet shopping services were frequently rejected as their affordances were accommodated in a particular paradigm of what was important in the moral order of the home. But gradually, a build of affordances facilitiated a radical shift in what was deemed important.

These shifts of paradigm in what are held to be deeply important values, remind us perhaps that these values are neither trivial in the sense that they are continually in flux, nor expressions of timeless moral principles. They are stable to the extent that they help frame the way in which what is afforded by information and communication technologies are comprehended. But they are capable of radical shifts through a process of gradually recomprehending and redescribing what is deeply important in the household. Upon such shifts, what was once afforded to the household through a technology becomes significantly different.

These shifts provide examples in a more prosaic domain of the redescriptions of moral orders that Macintyre holds to take place on a grander scale. More importantly, they provide a salutary warning to designers that the affordances of the devices they generate will not always be comprehended within the same paradigm. Approaches to investigating ethnographically the affordances of new devices and services might need to accommodate a longer time frame of investigation to allow these paradigm shifts to become apparent.

This thesis has considerable significance for the design of devices for households. Devices are frequently designed on the basis of particular models of how households might or might not use the devices, models generated by marketing data and by understandings of the preferences of potential customers. These studies may be more or less ethnographic in respect of the level of detail and theoretical orientation of the study. But the affordances of the devices, once designed, may have significant impact on the moral order of the home to the effect that the devices come to be used, or rejected, in ways quite contrary to the expectations of designers. Unless the ethnographic study is sufficiently long term, there is danger of it completing before this process of change has taken place and thereby misrepresenting the impact of the technology on the home. There is a case therefore for ethnographic pilot studies of new devices within households that explore the manner in which these shifts in the paradigms of the moral order can, and do take place.

References

Anderson, RJ (1994) "Representations and Requirements: The Value of Ethnography in System Design", *Human Computer Interaction*, Vol. 9, pp. 151–82.

Bannon, L and Schmidt, K (1991) "CSCW: Four Characters in Search of a Context", in JM Bowers and SD Benford (eds.), *Studies in Computer Supported Co-operative Work: Theory, Practice and Design*, Amsterdam: North Holland.

Benham, HC and Raymond, BC (1996) "Information Technology Adoption: Evidence from a Voice Mail Introduction", *Computer Personnel*, January.

Brassington, F and Pettitt, S (1997) *Principles of Marketing*, London: Pitman Publishing.

Engel, JF, Blackwell, RD and Miniard, PW (1990) *Consumer Behaviour*, Hinsdale, IL: Dryden.

Fergusson, D (1998) *Community Liberalism and Christian Ethics*, Cambridge: Cambridge University Press.

Frohlich, D, Chilton, K and Drew, P (1997) *Remote Homeplace Communication:What Is it Like and How might We Support it?*, Hewlett Packard Laboratories Technical Report, Bristol.

Giddens, A (1984) *The Constitution of Society*, Cambridge: Polity Press.

Holloway, R (2001) "Ethical Jazz", in T Bentley and D Stedman Jones (eds.), *The Moral Universe*, Demos Collection no 16, London: Demos.

Hopping, D (2000) "Technology in Retail", *Technology in Society*, Vol. 22, pp. 63–74.

Jones, K and Biasiotto, M (1999) "Internet Retailing: Current Hype or Future Reality?" *International Review of Retail and Consumer Research*, Vol. 9, No. 1, pp. 69–79.

Kuhn, T (1962) *The Structure of Scientific Revolutions*, Chicago: University of Chicago Press.

Liao, S, Pu Shao, Y, Wang, H and Chen, A (1999) "The Adoption of Virtual Banking: An Empirical Study", *International Journal of Information Management*, Vol. 19, pp. 63–74.

Macintyre, A (1981) *After Virtue*, London: Duckworth.

Macintyre, A (1998) *A Short History of Ethics*, London: Routledge (2nd edn).

Mateas, M, Salvador, T, Scholtz, J and Sorensen, D (1996) *Engineering Ethnography in the Home*, CHI 1996 Vancouver, BC: ACM Press.

Miller, D (1998) *A Theory of Shopping*, Cambridge: Polity Press.

Norman, DA (1998) *The Psychology of Every Day Things*, New York: Basic Books.

O'Hara, K and Perry, M (2001) "Understanding User-Centred Opportunities for Supporting Consumer Behaviour through Handheld and Ubiquitous Computing", under review.

Omar, O and Kent, A (2001) "International Airport Influences on Impulsive Shopping: Trait and Normative Approach", *International Journal of Retail and Distribution Management*, Vol. 29, No. 5, pp. 226–35.

Pahl, J (1989) *Money and Marriage*, London: Macmillan.

Parsons, T (1951) *The Social System*, London: Tavistock.

Rowley, J (2000) "Product Search in e-shopping: A Review and Research Propositions", *Journal of Consumer Marketing*, Vol. 17, No. 1, pp. 20–35.

Silverman, D (1970) *The Theory of Organisations*, London: Heinemann Educational.

Silverstone, R, Hirsch, E and Morley D (1992) "Information and Communication Technologies and the Moral Economy of the Household", in R Silverstone and E Hirsch (eds.), *Consuming Technologies: Media and Information in Domestic Spaces*, London: Routledge.

Underhill, P (2000) *Why We Buy: The Science of Shopping*, London: Texere.

Time as a Rare Commodity in Home Life

4

Lynne Hamill

4.1 Introduction

Why, in 2000–01, did more households in the UK have an Internet connection than a dishwasher? Dishwashers have been available in the UK for about half a century, yet only a quarter of UK households have one. In contrast, one-third of households have an Internet connection even though they have been available domestically for as little as 5 years or so (ONS, 2002a). Dishwashers save time, Internet connections absorb time. We're all supposed to be hard-pressed for time – so what's going on here?

Both money and time are in limited supply – both are scarce. We all say "I can't afford that" or "I haven't got time to do that". What we really mean is that we have limited money and limited time available and we prefer to buy or to do other things. Indeed, Juster and Stafford (1991) argue that time is the ultimate scarce resource because while incomes on average grow as economies grow and the standard of living rises, there are never more than 24 hours in a day. So how do households balance their financial and temporal budgets? Economics is concerned with the allocation of scarce resources (see Pearce, 1992; Hirschleifer, 1988) and this is therefore an economic question.

So we start by looking at the economics of the household. Then we look at the spread of domestic appliances in UK households; and take a close look at the impact of television and its associated technologies – video recorders (VCRs) and digital versatile disks (DVDs). The final section offers some lessons for producers of new domestic technology.

4.2 The Economics of the Household

A household can be regarded as both a producer and a consumer (e.g. Becker, 1965). While it consumes in order to maximise its utility, it also has to produce certain goods in order to survive, such as meals and clean clothes. Thus in the same way that firms combine capital and labour to produce output optimally, so do households. However, this similarity should not be overdone. While for firms, "optimality" means maximising profit, which is readily measurable by standard accountancy methods, for households, optimising means maximising utility (or welfare), which is a philosophical concept.

Nevertheless, by looking at "capital" as domestic technology, and "labour", time, we can see over the last century or so that there has been a continuing substitution of capital for labour within the household. Table 4.1 shows how the work done by Mrs Beeton's "maid of all work" in 1859 is done today (Beeton, 1859/1986): many of the jobs have been eliminated or considerably reduced directly by new domestic technology (or indirectly by clean fuels, better plumbing, easy care materials, better cleaning agents and so on). In other words, technology has been used to increase the productivity of those doing the housework. In effect, this "buys" time. It was at first the servant's time that was bought out, but increasingly it has "bought" time for all the family, particularly women. (In Britain in 1995, women still spent some two and a half hours a week on "domestic work" compared to about three-quarters of an hour for men (ONS, 1999).)

So how do people spend their time? Figure 4.1 shows that on average in Great Britain in 1995, people spent a third of their time sleeping. A fifth was spent working, studying and commuting. A further fifth was spent looking after themselves, their family and their homes, and much of this could be regarded as "unpaid work". (In general, the more time people spend in paid work, the less time spent in unpaid work: see, e.g. Lingsom, 1989). This left a quarter of people's time free.

In Chapter 3, John Strain refers to the negotiation that goes on within households to decide how both money and time are spent. Households decide how to allocate both their time and money in order to maximise

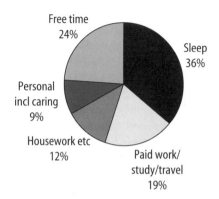

Figure 4.1 Time use in Great Britain, 1995.
Source: ONS (1997a).

Table 4.1. Tasks performed by a "maid-of-all-work" in 1859 and how they are done now

Task	Now	
	Labour-saving technology	Other labour-saving products
Open the shutters/windows		
Brush up kitchen range, . . . clear away ashes, clean hearth.	Central heating reduces need for cleaning.	Proprietary cleaning agents.
Polish with a leather the brightparts of the range	Extractor fan to reduce grease.	
Light the fire	Central heating.	
Put on the kettle, carry urn into dining-room to make the tea or coffee	Electric kettle, coffee machines. Timers	
Clean hearth and dust breakfast room	Central heating reduces need for cleaning. Vacuum cleaner	
Sweep hall, shake mats	Vacuum cleaner	
Clean doorstep, brass knockers/ handles polished up with leather		Proprietary cleaning agents.
Clean the boots		Easy care, wipe over
Cook the bacon, kidneys, fish, etc.	Microwave cooking	
Air bedrooms and beds		
Empty the slops from the bedrooms	Plumbing: inside, upstairs lavatories	
Make beds		Duvets
Wash clothes and linens	Washing machine and tumble drier	
Starch and iron		Easy care fabrics
Clean drawing-room	Clean fuels reduce need for cleaning. Vacuum cleaner	
Cook lunch	Timers. Microwave.	
Sweep up crumbs in the dining-room, . . .	Vacuum cleaner	
Sweep the hearth, and lightly dust the furniture	Central heating reduces need for cleaning	
Wash up dishes	Dishwasher	
Put away the lunch things		
Sweep dust and tidy the kitchen		
Needlework		
Shopping	Internet shopping	
Put on the kettle for tea	Electric kettle	
Clean knives		Stainless steel knives
Take in the tea		
Turn down the beds		
Fill water-jugs	Plumbing: upstairs/en-suite bathrooms	
. . . and bottles	Central heating and electric blankets	
Close the windows, and draws down the blinds.		
Clean up glasses, plates, etc. which have been used for the evening meal	Dishwasher	
Place wood near the fire, on the hob to dry	Central heating	
Lock and bolt the doors		

their utility. It can be seen as a joint decision: how much time is each member of the household willing to spend working for money? This will depend on their skills, the wages on offer etc. Becker (1965) argued that people will choose to allocate their time between paid work and other activities so that at the margin the extra unit of utility gained by work will equal that gained by leisure. The utility gained by work is, essentially, the wage earned. However, as real wages rise people will not necessarily work more, simply because of diminishing marginal utility. Marshall (1890/1961) pointed out that "the additional benefit which a person derives from a given increase in his stock of a thing diminishes with every increase in the stock that he already has". In other words, the more a person has, the less he values it. This holds for money just the same as for other commodities, so: "every increase in resources increases the price which he is willing to pay for any given benefit". Rising real incomes can therefore lead to a reduction in hours worked.

But this fine-tuning assumes that people can work the hours that suit them best. However, in practice, most people have to take what is offered by employers. In other words, the decision is not at the margin. They cannot fine-tune their labour supply decision, but have to decide, say, whether to work or not, or whether to work 40 hours or 20 hours. Mulgan and Wilkinson (1995) report that "there remains a serious mismatch between the hours people want to work and those available to them. Over 70 per cent of British workers working over 40 hours want to work less." Furthermore, Gershuny (1995) reports that "the better off part of British society really does want shorter working hours". There is little flexibility and this means that people have to look to other ways to balance their lives. So for instance, Mum can work 40 hours a week, but only if the household buys pre-prepared meals and labour-saving appliances. Thus households' labour supply decisions and consumption patterns are directly linked. The introduction of new, time-saving technology could potentially alter this balance between paid and unpaid work.

4.3 Technology in the Home

Over the last half century there have been significant changes in both work patterns and the domestic technology available. The main change in work patterns is that whereas in the 1950s the majority of British women with partners did not do paid work outside the home, now the majority do. Given this change in female work patterns, we would expect to observe a dramatic rise in ownership of labour-saving devices. Over the same period, the share of consumers' expenditure accounted for by durables (including cars) has doubled. All this would suggest that we should expect British houses to be full of time-saving durables.

But this has not happened. Bowden and Offer (1994) noted that "home entertainment appliances such as radio and television have diffused much faster than household and kitchen machines". In the UK in 2000–01, the top six consumer appliances were owned by over three-quarters of households. Of these, three were entertainment devices, time-users: televisions, video recorders and compact disk (CD) players. Indeed, of the twelve appliances listed in Table 4.2, only seven could be called time-savers.[3] The rest are time-users in that they are entertainment. This is consistent with the finding that in the UK, since 1998–99, expenditure on leisure goods and services has been the largest single item of household expenditure – accounting for 18 per cent (ONS, 2002a).

Table 4.2. Ownership of domestic appliances in the UK

	Percentage of households with appliance in 2000–01	Year introduced to UK households	Year reached 50% penetration	"Half life" (years)
TV[a]	99			
Black and white		1948	1958	10
Colour		1967	1976	9
Telephone	93	Pre-WW1	1975	60?
Washing machine	92	1934	1964	30
VCR	87	1979	1988	9
Microwave oven	84	Mid-late 1970s	1990	15?
CD player	77	Early 1980s	1995	15?
Tumble drier	53	1950	1994	44
Mobile phone	47	1990?	–	–
Home computer	44	Mid-late 1970s	–	–
Satellite/cable TV	40	1982	–	–
Internet connection	32	Mid-1990s	–	–
Dishwasher	25	1957	–	–

[a] 1996.
Sources: Bowden and Offer (1994), Office of Population Census and Surveys (1996), ONS (2002a).

It is well known that the rich are early adopters of new technology. Figure 4.2 shows the difference between the richest and poorest households in ownership of domestic appliances. For each appliance, the bottom of the line represents the proportion of the poorest households who own it and the top of the line, the proportion of the richest households. So, the longer the line, the greater the difference in ownership rates between rich and poor. The square represents the proportion of all households. So, for example, 83 per cent of the poorest households have telephones, as do 99 per cent of the richest and 93 per cent of all households.

---

[3] Phones – both fixed line and mobile – are counted as time savers. Bowden and Offer (1994) argue that the telephone is a time-saving device on the basis of the pattern of diffusion among households. Stehmann (1995) agrees, pointing out that although "the usage of a telephone is time-consuming in itself; on the other hand, the telephone is a time-saving means of communication compared with its substitutes: personal contact, letters etc."

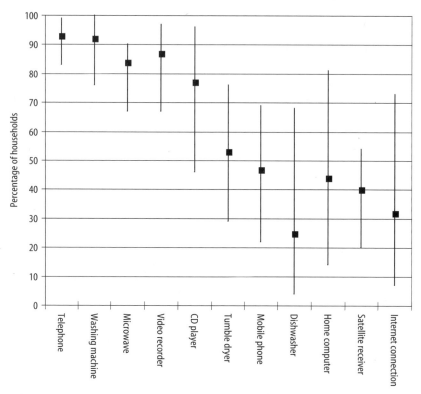

Figure 4.2 Ownership of consumer appliances, UK households, 2000–01: range by income decine.
Note: Figures are not available for televisions because they are almost universal (ONS, 1997b; 2002a).

Defining as "common" those appliances that are found in over about half of households, and dividing the appliances between time-savers and time-users yields a taxonomy shown in Table 4.3. This produces a host of questions, such as why are dishwashers common only in rich households?

Table 4.3. Ownership by type of domestic appliance: UK, 2000–01

Type of durable	Frequency of ownership	
	"Common" in all households	"Common" only in richest households
Time-savers	Telephones Washing machines Microwave ovens	Mobile phones Tumble dryers Dishwashers
Time-users	Television Video recorders CD players	Satellite receivers Home computers Internet connections

Richer households will substitute capital for labour – they buy domestic appliances to save time. To understand this, Douglas and Isherwood (1979) present the idea of "periodicities". They argue that "between households of different income levels, to be poor is to be periodicity-constrained in the processs of household management". In other words, the poor have to spend more time doing regular chores while the rich can afford new technology to free them. Consequently, a change in life-styles can be identified by a change in periodicities – or time use. Further, they propose that "Periodicities give a rough approximation to a major difference in the use between necessities and luxuries: future necessities in the present luxury class will be sets of goods with effective periodicity-relieving properties." Over time, as the economy grows and people become better off, things that were once only affordable by the rich spread to the rest of society.

However, according to the neoclassical economic model of the household, given a household's income, what the household actually buys depends on the costs of the goods and services available and the household's tastes.

There are several different types of costs that are relevant. Obviously, there is the price of the product itself. But the cost may not simply be the price tag. Domestic appliances are often bought on credit, which means that people pay significantly over the price tag. In addition, there may be running costs too: for instance, a home Internet connection requires expenditure on phone calls. Some require "durable complementary assets", such as CDs and VCRs (Shapiro and Varian, 1999). There are also the prices of alternatives, substitutes. Hiring a video or DVD is an alternative to going to the cinema for example. To understand the likely demand for domestic technology, these complements and substitutes need to be identified.

Of course, any pair of households with identical incomes and choices may make quite different expenditure decisions depending on their tastes. However, these tastes will be influenced by the decisions that they see other households making. This positive feedback on demand can operate in two basic ways. First, and most obvious, is the spread of those types of new technology that depend on "network economics". While the effect appears under a variety of guises such as "network externalities" and "demand-side economies of scale", the key point is that "the value of connecting to a network depends on the number of other people connected to it" (Shapiro and Varian, 1999). It is this phenomenon that gives rise to positive feedback – the bigger the network gets the greater the benefit of being connected to it. Clearly, the telephone is the classic example, and now, e-mail. The second source of positive feedback does not just apply to network technology. Douglas and Isherwood (1979) call it the "infectious disease model", the powerful force of wanting things one's friends and relations have, "keeping up with the Joneses". Not only does this apply to the decision on whether or not to buy a particular good, but also on the choice between competing goods. As Arthur (1990) points out, technologies typically improve as more people adopt them.

4.4 Case Study: TVs, VCRs and DVDs

TV is perhaps the ultimate time-using technology and TV has come to dominate our free time.

The BBC started TV broadcasts in 1936. Then, there were just "a few thousand in the London area" who could receive the transmissions. However, as transmission was stopped during the Second World War, it is probably more realistic to regard 1946 as the starting date. By 1948, only 0.3 per cent of households had TV sets but by 1958, it was 52 per cent. By the 1970s, over 90 per cent of UK households had TV sets and by 1996, 99 per cent of households had them. Colour television broadcasting started in the UK in 1967 and by 1979, 66 per cent of UK households had colour TVs (see Figure 4.3; BBC, 2002; Douglas and Isherwood, 1979; ONS, 1997b).

The drop in price has undoubtedly been a major factor in its widespread adoption. When first introduced in 1936, TV sets cost "as much as 100 guineas" (BBC, 2002) which is probably over £2,500 today. Now they can be bought for under £100.

However, there are also running costs. The use of a TV in the UK also requires a licence. The current fee (2002) for a colour TV is £112, more than the cost of a small set!

Since its introduction, there have been three major extensions of the technology:

- to manage TV viewing (VCRs and TiVo);
- to increase its range – more channels (with the arrival of satellite, cable and digital TV); and
- to allow people to use TVs to watch films delivered directly from equipment in their own homes (VCRs and DVDs).

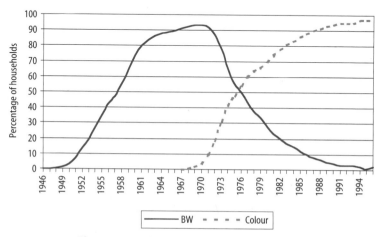

Figure 4.3 UK households with TVs: 1946–96.
Source: Bowden and Offer (1994), universal (ONS, 1997b; 2002a) and earlier reports.

Table 4.4. Use of free time in Great Britain, 1995

Activity		Hours per week
TV or radio		19
Socialising		11
Visiting friends	5	
Talking/socialising and telephoning friends	3	
Eating and drinking out	3	
Activities		9
Reading	3	
Walks and other recreation	2	
Hobbies, games and computing	2	
Sports participation	1	
Religious, political and other meetings	1	
Other		2
Total		42

Source: ONS (1996).

In 1981, Sharp observed that "Since TV has only been developed on a large scale in the last 30 years, it is apparent that it has had a very considerable impact on the use of leisure time". But as there are only 24 hours in the day, where has this time devoted to TV come from? What other activities were reduced? Table 4.4 shows how free time was used in Great Britain in 1995: watching TV or listening to radio dominates.

In 1939 the BBC conducted what was probably the earliest time budget study in the UK. Comparing the 1939 results, when there were virtually no TVs, with a similar study conducted in 1975, when almost every household had one, the BBC found that in 1975 people were home earlier than in 1950 and suggested that: "It could be that the working day has shortened, but there is also the possibility that the attraction of television has reduced the tendency to spend time 'doing nothing in particular'" (BBC, 1978). Juster and Stafford (1991) argue that in Japan and the United States, "reduced market work of adult men . . . was very possibly the consequence of new household technology rather than tax rates or rising income, and that possibly a further changing television technology (eg video) could have additional major impact on market activity".

And what about housework? The BBC (1978) study reported that:

> A more striking difference between the immediate post-War years and 1975 concerns the proportions engaged in "meals/domestic activities" during the evening. Even making allowances for possible differences in the definition of activities and in classification procedures, the amount spent in this way in 1948 seems to have been very much greater than in 1975. There are several plausible explanations for this. It is not unreasonable to assume that people did not make eating secondary to listening in the way that they now sometimes do to viewing, "convenience foods" were not then so common and "TV dinners" unknown and people probably lingered over their meals a good deal longer.

Bowden and Offer (1994) also suggest that the increasing time spent watching TV was taken from housework.

Vitalari et al. (1985) report two US studies that showed "television had a great impact on radio listening and magazine and book readership. There was an initial impact on attendance at spectator sports and a reasonably long-term impact on motion picture attendance." However, one of the studies quoted by Vitalari et al. (1985) found that "after three or four years of television adoption, families rearranged their lives and resumed their leisure activities".

In the UK, the introduction of TV had a major effect on cinema attendances. Between 1956 and 1960 average weekly cinema admissions in Great Britain fell by more than a half (from 21 million to 10 million). At the same time, the proportion of households with TVs rose from just under half to nearly three-quarters. By 1984, when TV was ubiquitous (and four-fifths of households had colour TVs), cinema attendance reached a low of just 1 million (CSO, 1990). Yet after 1985, cinema attendance recovered and is now at the same level as in the early 1970s (*The Times*, 1999; ONS, 2002b).

So time for watching TV was taken from every other type of activity – paid and unpaid work and other leisure activities. Thus the introduction of TV had a dramatic effect on peoples' use of time. Somehow a quarter of peoples' waking hours apparently became devoted to watching TV, although quite what that means in terms of activity is not so clear. Initially, the introduction of TV appears to have been given a special place in the daily routine, but over the years, it appears to have become absorbed into other patterns, reflecting what we call the rhythm of daily life.

In the early days of TV, transmissions were limited to certain hours and it appears that there was a tendency to watch TV in the sense of sitting down and focusing on it to the exclusion of other activities. Nowadays, TV is available around the clock. Furthermore in most cases, is free at the point of use. It is therefore not surprising to find that people are more likely to combine watching TV with other activities as this is entirely consistent with the economists' view of the use of time as the allocation of a scarce resource. People – implicitly or explicitly – will weigh up the costs and benefits of alternative ways of spending their time and choose the pattern that they most prefer. The financial cost of watching an extra programme on TV – certainly on the five analogue, terrestrial channels – is zero (ignoring the negligible cost of electricity). Economic theory tells us that the demand for such a service will be very large indeed as people will continue watching until the benefit to them falls to zero. However, there is also a cost in terms of peoples' time. They could spend that time doing something else. And, in a sense, this is precisely what they do, by combining "background TV" – when the set is on but not really watched – with other activities.

The little evidence available suggests that perhaps around two-thirds of the 25 or so hours a week people spend watching TV is what we might call "focused watching" as opposed to "background watching"; that translates to about 16 hours a week, or two hours a day.

In 1979, video cassette recorders arrived. By 1990, the UK Central Statistical Office commented that "video has been transformed from a specialised branch of communications technology to a mass domestic market" (CSO, 1990). At that time, about half the households in the UK had a VCR. By 2000–01, this had risen to 87 per cent. This rapid introduction of VCRs occurred despite the uncertainties caused by the initial competition between formats. "In consumer electronics, buyers are wary of products that are not yet popular, fearing they will pick a loser and be stranded with marginally valuable equipment." (Shapiro and Varian, 1999). Omerod (1999) argues that being new, people had little information on which to base their choice and hence chose what other people had chosen. Thus, once the VHS technology obtained a small lead, positive feedback meant that the demand snow-balled and Betamax, the alternative technology, died.

4.4.1 Managing TV Watching

Before VCRs arrived, people had to watch programmes when the TV companies chose, not when they chose. Hence the ability of VCRs to allow "time shifting" – recording TV programmes to watch later – would increase the appeal of TV watching and it was therefore expected that people would spend more time watching TV. However, the increase in TV watching was not very significant. Maguire and Butters (1994) report a study by *Which?* magazine in 1990 that found "almost one third of video owners almost never programme their machines". They also reported a study by Phillips that "only 70% of VCR owners actually used them to time shift, resulting in only 30 minutes extra viewing per week." Our own research has suggested that many programmes that are recorded are never actually watched because, people claimed, they did not have the time.

Despite this apparent lack of demand for time-shifting TV programmes, TiVo, described as "a set-top box, about the size of a VCR, which uses a hard disk drive instead of videotape to record programmes" is being promoted as a technology that "lets you put leisure time before television time – without missing any of your favourite programmes" (BSkyB, 2002).

The fact that time-shifting has not proved popular could in part be attributable to this distinction between "focused" and "background" watching. The amount of TV watched did not increase significantly through time-shifting; maybe by only half an hour a week – just 2 per cent of total TV watching time. However, people will only go to the trouble to record programmes that they think they want to see. In that case, the half an hour a week has to be set against the 16 hours or so of "focused" viewing, which implies an increase of about 3 per cent. Furthermore, the evidence that people fail to watch much of what they record suggests that even more time-shifting recording takes place than time-shifted viewing. The question is whether the failure to use VCRs for time-shifting is primarily due to the widely acknowledged problem of poor usability or because of fundamental aspects of people's behaviour.

4.4.2 Increasing the Range

The range of programmes offered through TV has been increased by the introduction of cable and satellite services. Satellite TV arrived in 1982, followed by the first cable service in 1984. Of course, cable has taken time to spread across the country. However, by 2000, satellite, cable and digital TV were found in 40 per cent of UK households (ONS, 2002a).

4.4.3 Films at Home

VCRs and DVDs are used for watching films on hired or bought tapes or disks. It is claimed that the DVD player "is one of the fastest growing consumer appliances ever". Launched in 1997, it is estimated that one-tenth of UK homes have one (BBC, 2001).

When VCRs arrived in the home after 1979, it was predicted that cinema attendances would continue their long-term decline because people could watch films at home. By 1986, 30 per cent of households had hired a tape during the previous week (CSO, 1990). Nevertheless, cinema attendance has risen quite dramatically. It appears that watching films at home is therefore not a substitute for going to the cinema. Of course, it does not provide the same experience for technical reasons: the visual impact and sound quality are much reduced on the small screen. But, there is also the attraction of "going out". Leisure inside the home and leisure outside the home are different types of activities that provide different benefits. Punie (1997) noted that there appears to be a tendency to divide leisure time between home and outside in the ratio 60:40. Perhaps the renewed popularity of the cinema represents people's efforts to re-establish this ratio?

4.5 Analysis

This chapter has looked at the adoption of new technology in the home from the perspective of time use. Because time is a scarce resource, the tools of economics can be applied to provide a different perspective on human behaviour from that provided by designers, engineers and other social sciences.

It is obvious that new technology in the home will compete for people's time alongside all the other demands on their time. Following Gronau (1977), the use of time can be divided into four categories:

- market work, which is paid;
- personal and biological maintenance, e.g. eating and sleeping;
- household work, which a third party could be paid to do;
- leisure where third-party production is conceptually impossible.

All four types of activity can be affected by the introduction of new domestic technology. For example, the arrival of TV not only changed how people spent their leisure time, but also seems to have affected the time they spent doing household chores and paid work too.

However, as we have seen, people have little flexibility in deciding how much time to spend on market work. Furthermore, most people need about eight hours sleep a day. So taking paid work and personal and biological maintenance as given, we can focus on the division between doing household chores and leisure. If people have enough money, they will pay to minimise the time and effort spent on chores and still have sufficient to spend to maximise their enjoyment of their leisure time. 'Twas ever thus for the leisured classes. However, most people are not in this fortunate position and have to choose between spending money to reduce the burden of chores and spending money to maximise the enjoyment of leisure. There is no point having lots of leisure time if there is no money available to do the things that you want to do. So you will not spend all your money on time-saving devices if that leaves you with no money to enjoy the time saved. Harper (2002) observes:

> One of the things that mobile phones have demonstrated is that people are pre-pared to pay a lot for idle chit-chat. When you move to command-and-control functions, such as opening doors remotely . . . We think that cost sensitivity will be much greater . . . I will pay anything to talk to my daughter. I'll even sacrifice my beer. But I don't want to sacrifice a couple of pints on Friday night just so that I can open my door.

On average, incomes grow over the years. If all the increase is spent on a time-saving device, there will be no extra income to spend on the extra free time released. Indeed the amount of income free to spend per leisure hour will fall because there will be more time to spread it over! The extra benefit from spending an extra pound on leisure is greater than the extra benefit to be obtained from spending an extra pound on reducing the burden of chores.

This also suggests that time is valued at different rates according to the activity: that people will be willing to spend more for an hour enjoying a CD than to save an hour spent washing up. Indeed, this is in line with findings elsewhere. Transport economists have discovered that the value people put on time varies according to "the context in which the savings or losses occur . . . the amount of time saved, whether the change related to increases or reductions in travel time, the duration of a trip, its purpose etc" (DETR, 2001).

This difference in values can also be related to the view of the household as simultaneously a producer and a consumer. The chores are equivalent to production and, just like firms, households will aim to minimise their costs while leisure is a consumption activity that contributes to households' utility, which they aim to maximise.

The use of domestic technology can be seen in the context of capital versus labour – of technology versus time. Some technology is used as an aid to production. They include the classic "white" goods. They are labour savers – time-savers – increasing the quantity of time available. In contrast, "brown" goods are time-users, typically offering entertainment, and therefore in a sense improving the quality of time. Of course, the traditional economic drivers of price and income are important, but it is also vital to take into account the effect on the household's time budget. It appears that people prefer to spend money on domestic technology to make better use of their leisure time than to save time on household chores.

The fact that for most people money is limited could therefore explain why people prefer to purchase time-using goods to time-saving goods. Hence, in 2000 more British households had Internet connections than had dishwashers.

Limited money, taken together with the inflexibility in the labour market and the need for personal and biological maintenance, it is not surprising that people's basic pattern of time use is, in the long term, surprisingly inflexible. This is underlined by the finding that new domestic technology may disrupt time-use patterns initially, but in the longer term people return to their original activities.

Forecasting the demand for novel domestic technology that has not yet been introduced is much more difficult than forecasting the demand for existing technology. Carey and Elton (1996) report that "the past century is littered with erroneous forecasts . . . some have seriously underestimated demand; most have overestimated demand". Having the best technology does not guarantee economic success; the marketplace does not guarantee that the best designed product wins. Back in 1934, Schumpeter pointed out: "The economic best and the technologically perfect need not, yet very often do, diverge, not only because of ignorance and indolence but because methods which are technologically inferior may still best fit the given economic conditions." Better understanding of how a new domestic appliance will fit into the time use patterns of households would improve the understanding of those "economic conditions" and thereby could improve the forecasting of take-up.

Unlike income, which for society as a whole increases over the years, the availability of time is constrained. While some new domestic technologies will save time, some will use time. This difference is key to understanding the rate at which they are adopted – in other words, their success in the marketplace.

References

Arthur, WB (1990) "Positive Feedbacks in the Economy", *Scientific American*, February, pp. 80–85.

BBC (British Broadcasting Corporation) (1978) *The People's Activities and Use of Time*, Audience Research Department.

BBC (2001) *Demand for DVDs rockets,* News Online, Business, 10 December (www.bbc.co.uk)

BBC (2002) www.bbc.co.uk/info/history/

Becker GS (1965) "A Theory of the Allocation of Time", *Economic Journal*, Vol. LXXXV, No. 299, pp. 493–517.

Beeton, I (1986) *Mrs Beeton's Book of Household Management*, London: Chancellor Press, first published 1859.

Bowden, S and Offer, A (1994) "Household Appliances and the Use of Time: The United States and Britain since the 1920s", *Economic History Review*, Vol. XLVLL, No. 4, pp. 725–48.

BskyB (2002) www.sky.com

Carey, J and Elton, M (1996) "Forecasting the Demand for New Consumer Services: Challenges and Alternatives" in RR Dhokalia, N Mundorf and N Dhokalia (eds.), *New Infotainment Technologies in the Home: Demand-Side Perspectives*, New Jersey: Lawrence Erlbaum Associates.

CSO (Central Statistical Office) (1990) *Social Trends 20*, London: HMSO.

DETR (Department of Environment, Transport and the Regions) (2001) *Transport Economics Note.*

Douglas, M and Isherwood, B (1979) *The World of Goods,* London: Routledge.

Gershuny, J, *Time Keynesianism*, The time squeeze, Demos, Issue 5, 1995

Gronau, R. (1977) "Leisure, Home Production, and Work-Revised Theory of the Allocation of Time Revisited", *Journal of Political Economy*, Vol. 5, No. 6, pp. 1099–123.

Harper, R (2002) "The Future Isn't all Bright for Orange's High Tech Home", *Sunday Times*, London, 12 May.

Hirschliffer, J (1988) *Price and Theory Applications*, Englewood Cliffs, NJ: Prentice Hall.

Juster, FT and Stafford, FP (1991) "The Allocation of Time: Empirical Findings, Behavioral Models and Problems of Measurement", *Journal of Economic Literature*, Vol. XXXIX, pp. 471–522.

Lingsom, S (1989) "Age and Behaviour", in *The Changing Use of Time, European Foundation for the Improvement of Living and Working Conditions*, Luxembourg: Office for the Official Publications of the European Communities.

Maguire, MC and Butters, LM (1994) "Usage Behaviour and Attitudes in Operating Home Electronic Equipment", in K Bjerg and K Borreby (eds.), *Home-Orientated Informatics, Telematics and Automation* (HOIT 94 Proceedings, University of Copenhagen), Copenhagen: Okios, pp. 397–401.

Marshall, A (1961) *Principles of Economics*, London: Macmillan, 8th edition; first published 1890.

Mulgan, G and Wilkinson, H (1995) "Well-being and Time: The Time Squeeze", *Demos*, Issue 5.

Omerod, P (1999) *Butterfly Economics: A New General Theory of Social and Economic Behaviour*, London: Faber and Faber.

ONS (Office for National Statistics) (1996) *Social Trends 26*, London: The Stationery Office.

ONS (1997a) *Social Trends 27*, London: The Stationery Office.

ONS (1997b) *Living in Britain: Preliminary Results from the 1996 General Household Survey*, London: The Stationery Office.

ONS (1999) *Social Trends 29*, London: The Stationery Office.

ONS (2002a) *Family Spending: A Report on the 2000–01 Family Expenditure Survey*, London: The Stationery Office.

ONS (2002b) www.statistics.gov.uk

Pearce, DW (1992) *Macmillan Dictionary of Modern Economics*, London: Macmillan Press Ltd.

Punie, Y (1997) "Created or constrained consumption? An assessment of the Demand for New Media technologies in the Home", *The Communication Review*, Vol. 2(2), pp. 179–205.

Schumpeter, JA (1983) *The Theory of Economic Development* (with introduction by JE Elliott); first published by Harvard University Press, 1934.

Shapiro, C and Varian, HR (1999) *Information Rules*, Boston: Havard Business School Press.

Sharp, C (1981) *The Economics of Time*, Oxford: Martin Roberston & Co Ltd.

Stehmann, O (1995) *Network Competition for European Telecommunications*, Oxford: Oxford University Press.

The Times (1999) "Star Wars Boosts Film Attendance", 25 August.

Vitalari, NP, Venkatesh, A and Gronhaug, K (1985) "Computing in the Home: Shifts in the Time Allocation Patterns of Households", *Communications of the ACM*, Vol. 28, No. 5, pp. 512–22.

Emotional Context and "Significancies" of Media

5

Sue Peters

5.1 Introduction

> It's about relationships, I think people have relationships with the Internet or their computers. It has something of them in there. You know, it has your name on the screen and you have your own favourite programmes, but the TV, I just don't have a connection with it, but that isn't me, that's somebody else giving me information (Female, 46).

Sometimes consumers do not use new technologies in the way that the manufacturers or content providers intend. The above example highlights one woman's acceptance of one new technology, the Internet, yet rejection of another, interactive television. While she felt she had a connection with her computer she did not feel she could actively use the television beyond information and entertainment. So what motivates consumers to buy devices and use them in certain ways while not using other technologies? This paper sets out to explore the reasons behind why new technologies have not necessarily been used in a way for which they were first designed. First, I will outline the often complex relationship between users or consumers and the devices which they use in everyday life. I will draw upon three devices, the PC (and Internet), the television (and interactive TV) and to a lesser extent the mobile phone. The main focus will centre on the relationship between the individual, family and these media or devices used within the household. I will then go on to explore generational differences between device and content use focusing on the concept of the media stream and the difference between background and foreground media. Finally I will look at the ethnography of social space drawing upon primary research using interpretive frameworks from Erving Goffman's concepts of regions and Thorstein Veblen's theory of conspicuous consumption to understand why certain devices are kept in certain spaces and how these spaces can interchange between private and social.

5.1.1 Background

I will use a number of frameworks for understanding consumer behaviour and device use but the basis of the findings stems from two empirical research projects conducted by me and my colleagues on behalf of research consultancy, Teleconomy. The first was an ethnographic study of 15 families incorporating a number of home visits to each household. The families represented a broad mix of demographic and socio-economic groups. A series of topic guides were developed to explore family media usage within the home and in particular their use of interactive television (iTV). In addition a small camera[4] was set up in the main living area which recorded the movements of the family within the room and also what they were watching on the television and how they interacted with the TV and each other. The second project was a qualitative and quantitative survey of over 500 families exploring device use within the home and the influence of the family on how different media were used. All quotes in italics are direct comments from the respondents.

5.2 Television

We are recently learning that television viewers do not gather to sit passively in order to watch the television. Up until the early 1980s there was an assumption that consumers behaved towards TV in the same way that they behaved when watching a film at the cinema. That is, they sat there obediently and watched the television as it was broadcast to them. This was a reasonable assumption as cinema looked superficially similar to television; after all, people sit and watch a screen in both cases. Sociological research (see Fiske, 1989; Morley 1986, 1992; Silverstone, 1994) exploded this myth of passive television consumption. For the first time researchers went into people's homes and observed their behaviour around TV. The results were unequivocal, and showed that people's behaviour around the TV could not be further from what had been imagined. Far from sitting passively and absorbing broadcasts, TV audiences play an active role, interacting with the medium in a way previously unheard of. People do sit there and just watch, at times, but, equally often, they talk over the TV, they discuss what is going on the TV with each other, they put the TV on in the background and ignore it, they wander in and out of the room, they carry on other activities in the room with the TV on, and they talk about the TV in other contexts (e.g. tea-break discussion about a programme on last night). In other words they dip in and out of the media, using it in an active way as part of the fabric of their social activity. Sociologists also have found that apart

---

[4] This was designed to be as "unobtrusive" as possible to capture the natural behaviour of the family and their relationship with and use of the television.

from relaxing, watching television provides a focal point for many families and acts as a means of "time-tabling" for family members. Charlotte Cornish, research manager at the Future Foundation, comments, 'In many households it is used as a way of organizing the week, with set soaps every day providing a routine for the whole family, which is very important . . . Television is used as a form of bonding for many families who all sit down to watch together' (Norton, 2000). This has implications for content providers who push their content on devices used in the main living area, and may also be pertinent for advertisers for analogue, digital and interactive TV who want to access the family or target children who may exert pester power over their parents, that is, children's increased voice in requesting purchases (Advertising Education Forum). Even though families do not necessarily sit down and watch television together as had traditionally been thought, my research shows that 88 per cent spend this shared time together in the main living area, this shared time mostly taking place in the evening (Teleconomy, 2002b). However, while they may not be passive viewers they equally do not appear to be active users especially where interactive services such as shopping are concerned.

Media pundits have laid claim to interactive digital television (iDTV or iTV) as a saviour of e-commerce in the business to consumer market. Indeed, digital television has enjoyed increasing penetration levels: at the end of 2001, there were over 8 million subscribers across satellite, cable and terrestrial platforms (Clawson, 2002). The Government has set conditions for coverage and take-up of digital TV services which must be met before analogue TV can be turned off. It has stated that it may be possible to meet these conditions and switch to digital TV between 2006–2010 (Parliamentary Office of Science and Technology, 2001). With users becoming accustomed to accessing more than just programmes via TV, with the popularity of electronic programme guides (EPGs), data rich programmes[5] and digital teletext, forecasts for the success of electronic transactions through the TV or t-commerce as it is becoming known as, have been optimistic. For example, Ovum predict that interactive services revenue will reach $62 billion by 2005, with the value of t-commerce at about $45 billion for that year (Ovum, 2001). However, retailers have been surprised to find that consumers are not using this medium to access their interactive sites to execute purchases. The current implementation of content does not necessarily equate with what users expect from the medium; as one respondent comments, *"Television is still some lowly beast over in the corner and it's an entertainment tool, not a practical tool."* Content providers therefore need to understand the relationship a television viewer has with the medium and, instead of withdrawing from this medium as a channel to market, address the reasons why consumers aren't using the device to purchase and act upon

---

[5] Such as BBC's Walking with Beasts.

this knowledge. Consumers are beginning to adjust their behaviour in relation to what interactive television can do and how it should be used but shopping isn't something that is associated with the TV (Teleconomy, 2001a).

CMP Media Inc. comment how pundits have long touted interactive television that would let viewers use their remotes for more than channel surfing. But many of the early pilots fizzled, leaving scepticism about the medium's future. Some commentators believe iTV will follow in the footsteps of cable, which was slow to take off but gained a following when it demonstrated real value in terms of content. However, some claim that due to the TV's relaxed personality people will never want to interact with the TV – they want to watch it and not have to think (Clawson, 2002).

Interactive TV does have primitive web-searching and surfing capabilities compared to the Internet through a PC but some analysts believe that the service enhances a medium with which people are comfortable. Philip Swann (2000) points to a very interesting question – why the TV rather than the personal computer? His answer – consumers are more comfortable with the TV than the PC, particularly at home. The TV conveys relaxation; the PC is associated with work, and "every good salesperson knows that the more relaxed your customer is, the greater your chances of making the sale" (p. 12). Swann continues to observe how Microsoft, which once believed that the PC would dominate all things electric, has changed its opinion and its marketing agenda and have developed enhanced TV software which is being installed in millions of cable and satellite set-top boxes.

Marshall McLuhan (1964) long ago distinguished between "hot" and "cool" media referring to the different sensory effects associated with media of higher or lower definition. Media such as print are said to be high-definition or "hot" as they allow for high sensory involvement on the part of the reader or listener whereas low-definition or "cool" media such as television require lower sensory involvement of the user because they require less effort on the part of the viewer. Interactive TV invites the viewer to become an active user but viewers need to make that shift in how they have traditionally used the TV set. Interactive TV demands a more active engagement of the viewer and requires the viewer to become more like a reader of the text. My research shows that many of the middle-aged television viewers were happy to use the TV as a cool medium, involving less sensory involvement than text-based media (Teleconomy, 2001a). To them the idea of using the *"lowly beast in the corner"* to conduct shopping was lazy and slovenly. Many of the older respondents also felt guilty at watching television, and many felt there was a psychological barrier to being actively involved with the television content beyond information and entertainment. In reality the television dominates many families' main living area and is used much more than anyone likes to admit. A *Radio Times* study of the nation's attitudes to television suggests British viewers are becoming increasingly dependent on their sets and, as in this research, many feel uncomfortable admitting so. Despite 40 per cent of homes leaving their TVs on for at least six hours a day, only 12 per cent of viewers admit they watch as much. And although reality TV and soaps are considered to be among the least important subjects on TV, viewing figures state the population still spends the most hours watching these shows (MORI, 2001a). Stephen Pile (2001) states that the BBC's findings are riddled with contradictions. Over 40 per cent claim that documentaries are their favourite type of programme, but the viewing figures are not so optimistic for this type of programme. My research at Teleconomy shows that the over 45s in particular are far more inclined to claim that they watch "serious" and informative programmes far more than they watch them in fact; in short, their rhetoric is more optimistic than behaviour. During focus groups the over 45s stated that they mainly watched documentaries and educational programmes when in actual fact they were as "guilty" of watching soap operas as the other age groups (Teleconomy, 2000). When this group did talk about using the TV they emphasised the wholesome aspect of it; when they stretched this to talk about entertainment it was almost confessional: *"in our family we're just as guilty as anybody else because we use the TV as a form of entertainment and also as a form of convenience because its easy to be entertained".*

This psychological barrier and lack of emotional attachment with the television are, among others, reasons for its failure as a transactional medium. Sometimes users do not know how to adopt a more active role as a user of the device and can only draw upon their experiences from interaction with another screen, the PC: *"that's because I'm thinking of it as a computer so I'm thinking that everything will automatically go down line and log and all the rest of it but of course it's not really interactive. It's almost like a dumb terminal, all it does is just throws you up pictures. There is no interaction. Whereas when you go on a computer and you press A and it'll take you somewhere else and it is very interactive. The TV doesn't interact with me."*

By its very nature, the television is a social screen, designed for entertainment and to coincide with the interests of millions of viewers. In the home environment as soon as more than one person watches the television there is a potential for conflict. Also, the conduit carries only one signal which can work in both directions. In another room the viewer is constrained by the viewing of the person who has the original source signal in their room, which, incidentally is normally in the living room. This also works in the other direction. Digital content can be linked to other rooms in the house and the same content can be viewed from many rooms. My research highlights that the respondents wanted to have the technology whereby different content could be accessed from different rooms in the house. Currently the only way to achieve this is by using a second set top box which would allow for different signals from a single connection.

The television should be seen as a medium which offers an enriched experience for the viewer and not necessarily as a means of accessing the Internet. Many of the respondents expected the television set to behave like a PC when they were accessing interactive content but felt uncomfortable because of the relationship they had with the set as a television. By drawing upon their experiences with a PC they felt they should be able to save the information they were accessing or be able to print it off: "*It should be like a computer, perhaps you could link up to the PC and you could download and you could copy and you could print*". Many respondents expressed a desire for a tangible copy of information or a receipt when purchasing through iTV. Perhaps this should be overcome by integrating content across media channels. For example, if the viewer is interested in a product advertised on television, such as a watch, they could register their interest by using the interactive services and a text message or an e-mail could be sent to them during or after their viewing. This way the information is redirected from the social screen, the television, to a more private screen, the mobile or PC. In the future (especially with location sensitive technology) marketers may be able to send the person a text message informing them they are outside a shop selling that watch. These devices should not be considered in isolation and their interrelatedness should influence the design of content. With "always on" technologies such as General Packet Radio Service (GPRS) to support mobile technology we will see an increased relationship between social and private media, as highlighted in the previous example.

Given this, it is clear that users need to be educated on how not only to use the device but also on how to perceive the content. We are beginning to see the PC-based Internet being used as a normal part of the shopping experience, yet many consumers still feel ill at ease when using the television for shopping, despite the fact that digital television penetration is becoming close to the Internet, with approximately 37 per cent of households having access to digital television and about 44 per cent having access to the Internet. Industry figures suggest it is set to supersede Internet access (Clawson, 2002). If users need to learn how to use the content more actively, content developers should take note of the relationship viewers have with their televisions. At the time of writing, interactive shopping services were being made to look like their counterpart Internet offerings, yet television viewers do not want to use this screen in the same way in which they would happily use the PC screen for the Internet. Acknowledging this reality, TV programme makers and advertisers have increasingly exploited our willingness to interact with TV in recent years. Television programmes have exploited links with other media, a notable example being *Big Brother*'s pioneering use of the Internet. Long before the Internet, however, TV programmes were using links with print media and the telephone (e.g. phone in voting, information lines etc.) to enhance the viewer experience. Television content should exploit what television is good at and what users are familiar with, notably audio and visual messages and narrative space and concision. We may see a move towards screens in the house rather than distinct devices for "set" tasks. These screens may offer a multitude of content but will be private or social depending on the nature of the activity.

In order for interactive content to be successful iTV needs to utilise its televisual qualities and the fact that viewers are very used to watching stories through the television. Interactive TV is not necessarily a medium in which users feel comfortable executing transactions. Commerce may never actually be accepted in the living room as Hulme (2002) highlights in his use of the following quote to explain how commerce compromises the sacredness of the living room. "It was both a shelter from the anxieties of modern life, a place of peace . . . and a shelter for those moral and spiritual values which the commercial spirit and the critical spirit were threatening to destroy" (Houghton, 1957).

Content providers should therefore not measure its success by actual transactions and incremental sales but by its role within the shopping experience. For example, my research at Teleconomy (2002a) showed that 63 per cent used iTV to look for products and services but only 3 per cent went on to execute the transaction via the television. Interactive TV can offer a different experience than the Internet but the inherent qualities of television itself need to be better considered. Consumer expectations of content are to televisual standards. In turn, for it to succeed as an interactive medium, it will need to become part of the user's everyday life, that is, a medium they will want to turn to automatically for such purposes as shopping, richer content, and communication.

Some analysts believe that the service enhances a medium with which people are comfortable, but I would stress that there needs to be consideration of the emotional significance of the set itself: *"If the TV was the computer with full Internet access, well, people will want to watch TV and others the Internet, how do you do that?"* So what is it about the PC that allows people to be comfortable enough to use the Internet and carry out all manner of tasks that they clearly don't feel as comfortable using the television? The next section will look at user behaviour with the PC and Internet and explore why people have a different relationship with it compared to the TV.

5.3 PC/Internet

Internet use via the PC is becoming embedded in everyday life for many people; 18 million people in the UK now have Internet access (BMRB, 2002). While the TV has been more familiar in everyday life it promotes passive use, while the PC-based Internet, by its very nature, promotes a more active use. I have often debunked the notion that people "surf" or "browse" the net (Teleconomy, 2000b); they are more directed in their use, even when using a search engine, most people have an idea of what it is they want to look for. Many people feel they have a connection or relationship with the PC that they cannot have with the TV: *"With the computer there's a connection. I don't know whether it's the touching of keys or whether it moves faster or it feels like I'm in control of it. There's no sense of that with television."* Recent research (Cahill, 2001) has shown that the number of URLs people use has decreased but rather than a decline in Internet use I argue this shows that people are displaying more purposeful behaviour in their Internet use (Teleconomy, 2002b).

Interactivity on the PC may involve other people online, such as in chatrooms, but the interactivity normally takes place with each person using one machine. In this sense interactivity depends on privacy and, unless it is set up to be social, the PC is an ideal medium for private content. This may be one reason for the apparent failure of the content that is successful on the PC, such as retailing or banking on interactive television. One respondent commented that while the TV was useful to look at his bank statement he carried out this activity when the kids had gone to bed and he was on his own. Research has shown that 50 per cent of people with access to interactive TV used this medium in their Christmas shopping experience to search for goods and services but only 9 per cent carried out the actual transaction using iTV (Teleconomy, 2002a). My research has shown that shopping using the PC was more purposeful and enjoyed higher numbers of transactions compared to iTV (Teleconomy, 2002b). In contrast with TV, the PC can be considered as a more private or semi-private device. If interactivity needs to be private and requires the interaction of one person then it makes sense for this content to be accessed via this medium. If, on the other hand, the interaction is more of a social activity, such as playing games or making a family purchasing decision, then it should be available on the social device or social screen. Current interaction on the social screen, the TV, requires interactivity from only one person, for example shopping using iTV, but consumers are beginning to expect devices to "talk to each other"; indeed, technologies such as bluetooth promise such a future. This expectation should be considered in the design of content, as users are almost expecting content from the social screen to be redirected on to another medium: *"I might look at it on the telly but then go and get some more info on the Internet, the trouble is that you have to fill in everything again."*

During my research I asked different age groups to visually depict the role of interactive TV in their lives. The younger groups (teenagers and pre-teens) all drew or cut out and stuck on a picture of a mobile phone next to the TV as if they were inviting the devices to talk to one another. Such behaviour was absent in the older groups' pictures. Consumers are becoming increasingly multi-referential in their shopping behaviour and rarely use a single medium or channel to follow a transaction through to the actual purchase. The PC-based Internet prompted more purposeful use, with 41 per cent using it to search and compare for goods and services and 58 per cent going on to purchase Christmas goods online (Teleconomy, 2002a). This finding is a considerable shift in behaviour and may mark a watershed for future Christmas shopping, whereby consumers are actually doing as much remote shopping, or even more, as they say they will be doing.

The purposefulness of Internet use may be one reason why we have witnessed a retreat of the PC from the main living area to a more private location. Research has shown that in January 2001 26 per cent of people located their PC in the main living area; currently this has dropped to 17 per cent with a predicted drop to 8 per cent in 12 months. The PC is retreating into the private domain with 63 per cent housing the PC in a bedroom or study (Teleconomy, 2002b). It is almost as if families were not quite sure of its role and therefore were unsure as to where it should reside. Similarly David Frohlich and Robert Kraut in Chapter 8 comment that finding a space in the home for the computer needs to fit in with the cultural norms surrounding different rooms and have highlighted the contentions in placing it in either a public or a private space.

The relationship that users have with PCs is then markedly different to the relationship they have with the television. The PC is viewed as more of a private medium in which directed tasks are achieved and purposeful behaviour displayed. The television promotes more of an opportunistic behaviour than the Internet, for example, advertising future programmes or programmes on another channel at the same time. The television is also often used in conjunction with another medium especially newspapers or TV guides and electronic programme guides. Users are accustomed to "flicking" or "channel surfing". In this sense TV viewing may be considered serendipitous. While many television viewers have set behaviours regarding scheduling they channel surf and hop around the content in a way not apparent in Internet behaviour. If users want to access content in the main living area yet do not want to interrupt their behaviour by leaving the room to boot up the PC, perhaps a better solution is using the mobile in conjunction with the TV? The next section will look at mobile user behaviour.

5.4 Mobile as Hybrid

I have already mentioned the desire of consumers to redirect content from the social onto more private devices. While the mobile is small and may not be the most appropriate device to perform the range of tasks performed on a PC it does represent a very private device and one which many users claim they can't be without. Perceptions of the mobile phone are already changing; the mobile is taking on a new meaning and has superseded its utility as a medium solely for voice telephony; it is increasingly perceived as a multi-purpose device. When the mobile phone first came into the market it was indeed just that – a phone that could be used when mobile, like a phone with its cord cut off (Teleconomy, 2002b). As mobile penetration levels have increased – industry commentators suggest that up to 75 per cent of UK adults now own a mobile phone (OFTEL 2002, MORI, 2001b) – so has the ability to do more with the device. The mobile can be used as a communicator through voice telephony, SMS text messaging and e-mail, an entertainment device through games, a data collection tool through WAP and GPRS, an alarm clock and an address book, as well as having other additional peripheral capabilities.

Along with its obvious communication capabilities the mobile is used by some as a tool to make a statement about the user, to display conspicuous consumption – the term for spending for the sake of prestige (Veblen, 1956). Veblen argued that self-esteem had become directly linked to the possession of material goods, something that can be seen in the use of mobile phones. The physical design and presence of the mobile has alternate meanings that go beyond its utility, such as status, social connection and even popularity. It is possible that teen mobile owners view their devices as extensions of themselves and their personalities (Alexander, 2000). Sadie Plant has made some interesting observations on types of mobile users, relating them to different types of bird. The swift, a bird that is elegant in flight and is always flying, always mobile, very rarely stops and may represent the mobile user who uses their phone away from the social space and is quite unobtrusive. In contrast is the peacock who uses the mobile as a matter of public display, may represent those users who happily display their mobile activities in front of others (Motorola, 2001).

Recent figures released from the Home Office indicate the mobile's desirability as a target for theft; 1.02 million sets were stolen over a 12-month period in 2000–2001. In 23 per cent of incidents overall victims were using their phone or had it on display when it was targeted (Harrington and Mayhew, 2001).

In my previous research it emerged that the younger groups used their mobile in many social spaces with a view to displaying their patterns and character of social relationships, or as a way of displaying who they were popular with. By tapping into their displaced networks through phoning or texting, users are making a statement that they have connections beyond their immediate surroundings (Hulme and Peters, 2002). By connecting to this network which is geographically displaced the user is showing that he or she is in what Goffman (1971, p. 19) would refer to as a "with", defined as "A party of more than one whose members are perceived to be 'together'". The user is signalling that he or she is not alone, and thus the mobile used as a "with" is a sign of membership and belonging as well as a practical means of achieving security. One characteristic of a "with" is civil inattention, again reinforcing a message to anyone in the surrounding environment, "don't talk to me" and "I am connected". The mobile is a hybrid in that among other things, it is both a private and public device, a means of maintaining privacy as well as demonstrating connectedness. It has similarities and differences with the use of the PC and iTV. In terms of a communicator the mobile can be likened to the PC-based Internet, certainly it is becoming increasingly common to access e-mails via the mobile and both devices are used in a private way. However, due to its size, portability and technical limitations the mentality of mobile use is markedly different to that of the PC. Previously I have argued that users are used to viewing tele-visual content over the TV and are only just beginning to learn that the set can be used more interactively. The opposite can be said for a mobile phone – its use is extremely interactive but users are beginning to learn to use it for visuals. This will become more prevalent with the advent of multimedia messaging which will enable pictures and even photographs to be sent, just as text messages are sent now. Additionally, 3G will, in theory, deliver multimedia, high-speed data and even real-time video images to our handsets when we are on the move (*Teleconomy, Quarterly Review*, 2001c). But, the mobile will not be the most appropriate device to watch films and in this respect will not be similar to how we use the TV. Perhaps we should be looking more towards using the devices in conjunction with one another. Certainly, many of the younger respondents showed a propensity to use their mobiles in front of the TV.

5.5 Media Streaming

Previous research (Peters and Hulme, 2001) has highlighted the fact that we live in a "media stream". Certainly in a work setting and in the home there is often a selection of media being used or available for use, the PC, radio, TV, mobile phone and newspapers. Such media, as Du Gay et al. (1997) would suggest, are often deployed for routine, everyday use. However, what is interesting is how different age groups use the devices. The younger groups' behaviour, that of teenagers and 18–25 year olds, was markedly different to the older groups. The younger groups preferred to run forms of media such as television, radio or permanently connected mobile phones concurrently in "background" mode, picking out items of interest from this "flow" of content. Among a number of significant findings the research indicated that in the age group 18–25 the home computer was often regarded as "outside" their normal range of behaviours. Stand-alone computers were seen to be outside this "flow" requiring switching on, loss of time and changes to behaviour which were considered intrusive: "*It's boring, you have to sit in a room and it's just you and your computer.*" The Internet was seen as a solitary, lonely place and also required a change in this group's behaviour, because it was often situated in a study and required time to "boot up" (Peters and Hulme, 2001).

The older age groups foreground their media more consciously and selectively. For example, when they entered a room to watch TV they sat and watched the programme. Similarly, when they left the room they switched the TV off. The younger groups resolutely left the television or radio on so that when they next entered the room there would be noise, *"I actually turn the radio on every time I walk in a room so that's probably companionship, I can't stand the silence."* The use of Walkman/Discmans demonstrates the same phenomenon: a mobile environment, Du Gay et al. (1997) argue that this portable device is part of the required equipment of the modern "nomad" – the self-sufficient urban voyager who operates within a self-enclosed, self-imposed bubble of sound.

We have only to enter a teenager's bedroom to enter a cacophony of noise. Quite often they may have the TV on and a CD playing; perhaps they are attempting their homework while also text messaging. What is interesting is that this younger age group are particularly adept at handling this noise and at having an ongoing background of media in use, actively or passively. One teenager commented, *"Anything for background noise. Because I don't like silence."* Mobile use is considered background by some respondents who are able to text message while riding a bike and even driving! (Teleconomy, 2002b). In many cases these images and noises run together, for example, the CD or radio plays and the television is on simultaneously. From this "stream" of media they pick out items of interest. In other words they are constantly available for persuasion or contact. The under-25 group were willing to be "contactable": *"I never turn mine off"*, whereas the older groups were more selective: *"I turn mine off at night and I think I only really have it in case of an emergency."* This leads to such statements, in the context of the mobile phone, as *"when I lost my mobile I was like panic stricken"*. In short the 18–25 group looked to access media content or "items" from among a flow which was ever present (they simply turn devices on to break up silences). Individual "items" tended to emerge from this flow. This phenomenon indicates a behavioural adaptation designed to gain maximum benefit for minimum effort from our media saturated lives.

This generational difference in media use should also be a consideration for manufacturers and content providers. However, the challenge is to know who is using the media and when. For example, recent empirical research has shown teenagers to be remarkably cunning in their use of mobile phones within the household. Quite often, when they have no credit left on their pre-paid mobile phones, they use their parents' mobile to send text messages and make calls. While the network providers have the user's details as the Parent and may build up a customer profile accordingly, all of a sudden this is challenged by the children's exploitation of the device for their own purposes (Teleconomy, 2002c).

5.6 Routine Activity

To look at how a technology may be adopted within the home I want to draw upon the work by Felson (1993) on routine activities. Routine activities theory stresses both the repeated, habitual character of much human action and the ways in which this changes over space and time. Routine activities stresses the importance of understanding interaction in space and time and within the material environment. Although Felson was writing about the distribution and prevention of crime this can be applied to events in the social world other than crimes; all such events occur within structured and routinised ways of using time, inhabiting places, and interacting with other people and with the material environment of things. This may be one reason for the increase of crime in mobile phone theft as mentioned earlier. If habitual behaviour changes over time and space so too does the use of technology and devices. However, in order for a new technology to be accepted and used, particularly in the home environment, it must fit in with current activity. Frohlich and Kraut note in Chapter 8 that routines develop over time and these routines develop slowly when a new technology comes into the home. The difference between *routinising* Internet behaviour and iTV use lies in the hardware and associations with the device. The TV existed in most people's houses before the advent of iTV whereas the PC did not have these levels of penetration (and still does not) before home users adopted the Internet. It is the content of iTV which changes the use of the television from passive to active whereas, arguably, the PC inherently prompts an active use. Interactive television is an example of how user behaviour needs to change but not everyone is open to the idea of using it actively: *"I don't think probably for my generation TV will ever be anything more than just switch on, switch off sort of, you know switch on telly to watch something mindless. I'm not sure there'll ever be that relationship as there is with computers."* In Chapter 9 Cheverst et al. identify a "technology push" approach (Moran, 1993) in studying homelife and design processes overlook the people who use it on a day-to-day basis (Tweed and Quigley, 2000) and the need to understand how technologies fit into daily routines. Similar to Cheverst et al.'s comments on technology needing technical, social and ethical concern I would argue that content for iTV needs to be thought of in the same way.

It follows from this analysis that it is a mistake to expect that new technological devices and content will – or indeed can – be adopted quickly and without a period of time for reflection, adjustment and adaptation to new routines. "We assimilate new experiences by placing them in the context of a familiar, reliable construction of reality. This structure in turn rests not only on the regularity of events themselves, but on the continuity of their meaning" (Marris, 1974, p. 6). New technologies have to have something people want: for example, iTV may have a particular attraction for families with young children who are more restricted than families with older children or empty nesters. Many families with young children expressed a desire to use iTV for basic shopping needs. Restrictions on their mobility and independence through having young children were noticeable, and iTV was seen as a quick and easy solution. This was particularly prevalent for one cable user who wanted to be able to shop while also not disturbing the children's viewing through split screens, that is, to use iTV so that both needs were satisfied (Teleconomy, 2001a).

I am advancing an argument for the integration of iTV that is close to that of Frohlich and Kraut (this volume) on integrating the computer into the home in that it depends on how it is or will be used and shared by the family members. The TV may be ubiquitous, with 97 per cent of the population having access to it (Teleconomy, 2002b), but such penetration does not guarantee that iTV will automatically be widely adopted. Consumers are accustomed to a passive consumption of televisual content not an active involvement with it as a text-based or shopping service.

5.7 Regions and Blurring

It may be useful to think about the devices in terms of their socialness and privateness. The television is largely a social screen and as such, interacting on the TV causes problems and risks conflict; one father of three commented that *"war has been known to break out over the remote control"*. The PC may be considered a private, or at least semi-private medium and the mobile the most private of all. Du Gay et al. note that the division between public and private space is both material – in that it denotes an opposition between physical spaces and symbolic, in that these spaces are made to signify different things in relation to each other, the public signifying the universal, the collective and the rational, and the private, the emotional and personal. Likewise the space in which these social and personal devices are used may also be categorised by their public and private nature (Du Gay et al. 1997). While the mobile is a very private device it can also be used as a social device.

Du Gay et al. comment about the material and symbolic division of public and private space. "Material" because it denotes an opposition between physical spaces signifying the universal and "symbolic" because these spaces signify different things in relation to each other, which are more emotional and personal.

Another approach to understanding the relationship and the emotional context and significancies of media is by using Goffman's (1959) concept of regions. A region may be defined as any place that is bounded to some degree by barriers of perception. Regions vary in the degree to which they are bounded and according to the media of communication in which the barriers of perception occur. The "front region" is the place where the performance is given, and the "back region" or "back stage" is where the suppressed facts make an appearance. "A back region or backstage may be defined as a place, relative to a given performance, where the impression fostered by the performance is knowingly contradicted as a matter of course" (Goffman, 1959, p. 114). The back region is where action occurs that is related to the performance but inconsistent with the appearance fostered by the performance. The home can be easily categorised by regions by looking at the role of different rooms and how they are used – for example, the living room as a front region and the bathroom as a back region. As a front region the living room is often the area which houses status objects indicating social status and esteem, both objects relating to one's self esteem and collective objects, artefacts representing ties with groups outside the family (Riggins, 1994). In contrast, the back region would house stigma items. Riggins notes that living rooms contain relatively few stigma objects; more can be found in the private area of the home, notably bathrooms and bedrooms (p. 112). Given the role of the regions in the home it is important to think about where devices are situated within the home and therefore within these regions. However, while there is a tendency for a region to become identified as the front region or back region of a performance with which it is regularly associated, still there are many regions which function at one time and in one sense as a front and at another time and in another sense a back region (Goffman, 1959, p. 127). In this sense the living room may take on the role of both back and front region and can quickly change its role according to the context of use. The family together watching television cause the living room to act as a back region. But as soon as a guest, or "intruder", as Goffman would suggest, enters this environment the room will revert to its role as a front region. The use of the TV in the back region may be different to its use when the same space becomes a front region. Goffman comments that spaces can make this shift when a stranger enters, and, particularly as back regions are out of bounds to members of the audience, a shift to the role of front region is necessary in order for the right tone of formality to prevail.

Goffman suggests that telephones are "sequested" in the back region so that they can be used privately. Obviously, Goffman was writing before mobile phones existed, but this observation proves interesting. On one hand, some mobile users do indeed sequest their phones in an alternative environment, away from their company to a private location, whereas others actively display their phones and use them in the presence of others.

Many mobile users display signs of civil inattention to create, in effect, a back region around themselves. Within the home, the use of mobiles varies interestingly. Teenagers adopt this private device and enjoy its use within the most private locations of all within households, in the bedroom specifically, and they have also been known to text from the bath (Teleconomy, 2000). Texting allows for a blurring of regions – while in the main living area texting can take place quietly, and need not upset other people in the front region, yet there are occasions when texting seems essentially a back region activity.

5.8 Conclusions

It is interesting to look at the ethnography of social space in order to understand the relationship users have with devices, and thus to think about the implications for content providers and also for device manufacturers. The earlier example of the retreat of the PC from the social domain back into the private arena should cause us to think about how we are to understand the relationship between the user and the device. One understanding may be to subvert Veblen's notion of conspicuous consumption and think about the role of inconspicuous consumption – consumption behaviour that belongs in Goffman's back regions. Devices such as the television tend to dominate the front region and as such may be seen as items of conspicuous consumption. During my research at Teleconomy nearly a quarter of respondents claimed they would upgrade or replace the TV within the next 4 months to a year and a further 14 per cent stated they would purchase digital or iTV within the next 1–2 years (Teleconomy, 2002b). The mobile, on the other hand, transcends regions as it is actively used in both front and back regions, but is also used as a display of conspicuous consumption. Just over half of today's 7–16 year olds in the UK own a mobile phone and in consequence, handset upgrades are becoming ever more common: a quarter (24 per cent) of young mobile phone owners are now on at least their third handset. "In keeping with the way in which their mobile handset communicates who they are to the outside world, 46 per cent have changed the cover of their mobile handset and 45 per cent have changed their ring tone" (*Daily Research News*, 2001). In contrast to where the television and mobile are used, the PC is located within the back region of the home, most notably in the study or a bedroom. As such it may be seen as an item of inconspicuous consumption, as it is not on display, as the television is, in the front region. Aside from the Apple Macintosh, the history of the PC's design and look has been fairly uneventful; it has remained relatively unchanged over the last two decades. Coupled with the behaviour PC use promotes, that of purposefulness and relative privacy, the actual design and look of the PC may be a reason for its location in the back region and its function as a medium for inconspicuous consumption.

Within the home environment there will not be one "killer" device which will be used in isolation from others, but rather consumers will use devices in ways which reflect the relationship they have with them. As convergence of channels progresses, so the use of devices may fragment as people develop preferences for certain media for certain tasks. How consumers relate to the various media can be understood in terms of routine activities; new technology needs to "fit in" with people's habits and lives within the home. With 88 per cent of people stating they meet in the living room or lounge, and 64 per cent meeting there in the evening, the living room is indeed that, living. Multiple devices are located in this area and many are used simultaneously, for example, 63 per cent use the landline in the lounge and 43 per cent their mobile. When asked about the amount of time the family spent together, over a third said that this time had increased from a year ago and a further 22 per cent said that it would continue to increase over the next year. Taking into consideration the importance we have seen placed on family devices, particularly the TV and digital TV, families seem to be venturing back into the main living space to spend more time together (Teleconomy, 2002b). Conspicuous consumption is more likely to occur in the main living area and inconspicuous consumption in the back regions where guests or strangers are less likely to intrude. The content accessed on these devices according to their locations in the home needs to tie in with the expectations users have of the different devices. As technology moves on, the same content will begin to be accessible on a number of devices, but, if it is to have the desired impact, users will need to relearn how to use the devices and alter their expectations of "one device for one task" and their associations of certain types of content with a particular medium. In the meantime it is useful to look at the relationship users have with certain devices, the emotional context of their use, and the ethnography of the space which these devices occupy, to understand the "significancies" of media and to inform the design of new media technologies.

References

Advertising Education Forum, *Advertising, Parents and "Pester Power"*, http://www.aeforum.org/issues/Children_pester_power.html

Alexander, PS (2000) "Teens and Mobile Phones Growing-up Together: Understanding the Reciprocal Influences on the Development of Identity", Submission for the Wireless World Workshop, University of Surrey.

BMRB's Internet Monitor (2002) http://www.bmrb.co.uk/interactive/intenetmonitor.htm

Cahill, J (2001) *Home Internet Use Continues to Grow in the UK*, 8 November, http://uk.netvalue.com

Clawson, T (2002) "Boxing Clever", *E-business*, January.

Daily Research News (2001) "NOP's M-Kids Outlines Youth and Mobiles", 11 December, http://www.mrweb.co.uk/

Du Gay, P et al. (1997) *Doing Cultural Studies: The Story of the Sony Walkman*, Buckingham: Open University Press.

Felson, M (1993) *Crime and Everyday Life*, London: Pine Forge.

Fiske, J (1989) *Television Culture*, London: Routledge.

Goffman, E (1959) *The Presentation of the Self in Everyday Life*, London: Penguin.

Goffman, E (1971) *Relations in Public*, New York: Harper & Raw.

Harrington, V and Mayhew, P (2001) *Mobile Phone Theft,* Home Office Research Study 233, London: Home Office.

Hulme, M (2002) "The Living Theatre", Conference paper for the *Interactive Home,* London Business School 11 July 2002.

Hulme, M and Peters, S (2002) "Rethinking Networks: Identities and Connectivity in the Global Age", Submission for "Absent Presence: Localities, Globalities and Method", Helsinki, 10–12 April.

Houghton, WE (1957) *The Victorian Frame of Mind, 1830–1870,* New Haven: Yale University Press.

Marris, P (1974) *Loss and Change,* London: Routledge.

McLuhan, M (1964) *Understanding Media,* London: Routledge.

Moran, R (1993) "The Electronic Home: Social and Spatial Aspects", Report of the EC's European Foundation for the Improvement of Living and Working Conditions, Luxembourg: Office for Official Publications of the European Communities.

MORI (2001a) *Radio Times View of the Nation Television Survey,* 28 August, http://www.mori.com/polls/2001/rt010828.shtml

MORI (2001b) http://www.mori.com/polls/2001/rt010828.shtml

Morley, D (1986) "Family Television: Cultural Power and Domestic Leisure", *Comedia* Series 36.

Morley, D (1992) *Television, Audiences and Cultural Studies,* London: Routledge.

Motorola (2001) *Talk is Cheap,* http://www.motorola.com/mot/documents/0,1028,301,00.doc

Net Value (2001) *Home Internet Use Continues to Grow in the UK,* 7 November, http://uk.netvalue.com

Norton, C (2000) "Two in Ten Watch TV 36 Hours Every Week", *The Independent,* 19 May.

OFTEL (2002) "Consumers' Use of Mobile Telephony Q8", February, www.oftel.gov.uk

Ovum (2001) "Digital TV – Partner or Foe in the Internet World?" 26 February, www.ovum.com

Parliamentary Office of Science and Technology (2001) "e is for everything?", Report, 17 December, London.

Peters, S and Hulme, M (2001) "Me, My Phone and I: The Role of the Mobile Phone", Submission for mobile workshop, CHI, 2001, Seattle, USA.

Pile, S (2001) *View of the Nation Television Survey,* http://www.radiotimes. beeb.com/content/webclub/view_of_the_nation/

Riggins, SH (1994) "Fieldwork in the Living Room: An Autoethnographic Essay", in SH Riggins, *The Socialness of Things: Essays on the Socio-semiotics of Objects,* New York: Mouton de Gruyter.

Silverstone, R (1994) *Television and Everyday Life,* Routledge: London.

Swann, P (2000) *The Future of Interactive Television: TV dot COM,* New York: TV Books.

Teleconomy (2000) *Ubiquity: Which Message, Which Medium?* Teleconomy report, Lancaster.

Teleconomy (2001a) *Interactive Television: Exploring User Behaviour,* Syndicate project, Lancaster.

Teleconomy (2001b) *Shopping On and Off the Web: An Analysis of How the Internet Features as Part of the Purchase Process,* Teleconomy report, Lancaster.

Teleconomy (2001c) Quarterly Review, March, Teleconomy Report, Lancaster.

Teleconomy (2002a) *Christmas Shopping: An Analysis of the Role Played by Virtual and Physical Channels in Christmas Shopping Hehaviour,* December 2001/January 2002, Lancaster.

Teleconomy (2002b) *Devices in the Home: A Study into the Behaviour and Use of Media and the Influence of the Family Within the Home,* Teleconomy Report, Lancaster.

Teleconomy (2002c) *The Mobile Effect: Change in Media Use,* Teleconomy Report, Lancaster.

Tweed, C and Quigley, G (2000) *The Design and Technological Feasibility of Home Systems for the Elderly,* Belfast: The Queens University.

Veblen, T (1956) *The Theory of the Leisure Class,* London: Unwin Books.

Part 2
Designing for the Home

Paper-mail in the Home of the 21st Century

6

Richard Harper and Brian Shatwell

6.1 Background

In the 1960s, the British Government told the Post Office that it would be out of business by the middle of the following decade. Telephones and thereafter fax would undermine the need for written paper-based communications. The Post Office was to prepare for bankruptcy. Forty years later, The Royal Mail, as it is now known, delivers more letters than ever before. Why is this? How can assertions about the future of paper-mail be so wrong? Why is the business continuing to expand?

These questions have become all the more pertinent at the start of the 21st century when the impact of the "Digital Age" is expected to be greatest. Will paper bills be replaced by electronic bill payment and presentment (EBPP)? Will the much-cherished handwritten letter be replaced by e-mail? And will direct marketing sales literature be delivered to people's Internet addresses rather than to their letter boxes?

It is no wonder, therefore, that numerous attempts to predict the future of paper-mail have been commissioned in the past few years. In Silicon Valley, for example, the *Institute of the Future* has been funded to look at the future of mail at a global level (http://www.iftf.org), while in Europe, various mail companies have funded similar though smaller scale investigations (e.g., Nikoli, 1998; Coopers and Lybrand, 1996). The same is happening in Japan (e.g. Izutsu and Yamaura, 1997).

All of these studies have themes in common. In particular, they include examination of the increasing uptake of the home PC, the widening of access to the Internet and the ever greater willingness of companies to offer EBPP. In combination, these factors are said to provide the basis for the substitution of paper-mail with digital alternatives.

This existing research has also highlighted certain cultural factors, such as the resistance to home PCs within certain lower income socioeconomic groups in the USA. Here, disposable income is utilised in quite different ways from higher income families, with an emphasis on entertainment (such as with digital TV) and much less on infotainment, as is perceived to be provided with the Internet. In Scandinavia, there is broader acceptance of computer technologies in the home, and thus it is predicted substitution will occur more quickly there than anywhere else. Finally, this research has also uncovered some attitudinal preferences for "quality paper" in mail, which has suggested that mail recipients view the quality of paper as an indicator of the quality of the sender. Colour and envelope design are obviously factors here as well.

The "substitution argument", as it is often known, has turned out to be very useful, especially given that it can use basic socio-economic indicators, such as per capita income, to specify the future rates of substitution. Yet, there are some doubts about the long-term accuracy of this research since the predictions are not being borne out. As with the predictions that paper-mail would disappear by the end of the 1970s, so now there is doubt as to whether these more recent analyses will turn out to be accurate.

6.1.1 A Conceptual Approach that Might Provide Answers

This kind of research focusing on the substitution of paper-mail by digital technologies brings to mind similar predictions about the future of paper in office environments. At least as early as the mid-1970s, the "paperless office" was becoming a popular catchphrase, and many pundits prophesized it was merely a matter of time before it became a reality. Investment rates in technology and more user-friendly technology were just a couple of factors that were believed to ensure the eventual paperlessness of offices. But paperless offices never appeared (Sellen and Harper, 2001).

The failure of that revolution – and indeed the continuing failure of paperless offices to materialise – was typically explained (and often still is) by reference to what was called "cultural factors". According to this view, paper continues to be used because those generations of people who were brought up with paper documents find it difficult to move towards screen-based documents and new technological tools. As this generation gradually retires, so digital documents will replace paper.

As it happens, investigation of this thesis indicates that there is very little relationship between age cohort and preference for paper. Instead, research has suggested that the reason why paper continues to be so important in office life has to do with its "interactional properties", or those physical aspects of paper which shape the ways in which it can be used in a whole range of different kinds of tasks (Sellen and Harper, 1997; 2001). These may be thought of as the *affordances* of paper.

It is worth mentioning what some of these affordances might be since the parallels between these and the affordances of paper-mail would seem intuitively obvious. In office environments, paper affords ease of marking. This turns out to be important when people are trying to review the contents of a document, allowing them to write and comment on the text as they read. One might imagine that similarly recipients of paper-mail would utilise the same affordance when for example, they tick or cross out items on a bank statement when they "balance the books".

Studies in offices show that paper also affords flexible cross-referencing between multiple documents, allowing users to spread pages out in physical space and to read and write "across" documents. This is important when people are trying to compare and contrast between documents or extract and integrate information across documents (all of which are common office activities). Similarly, one can imagine that when someone is balancing the books as just mentioned, such cross-referencing of information may also occur and that therefore this particular affordance of paper would also offer benefits alongside the ability to mark up and annotate. Items delivered through the post that may afford this would include not only bank statements, but such things as car insurance certificates that need to be checked against other paper documents, and so on.

It is not so easy to see how other affordances of paper that are important in offices provide benefits in the home, however. For example, paper also affords complex, two-handed navigation within and between documents. This enables office workers – particularly knowledge workers – to more effectively get to grips with the structure of a document by allowing them to quickly flick through and feel "where they are". One cannot readily imagine how people in home settings need to satisfy the same requirement. Of course they may do. But this is an empirical question. Similarly, in offices paper affords people opportunities to interact and communicate. Occasionally, for example, they may print-out "hard copy" so as to justify hand delivery of an important document to their boss, rather than e-mail it. This may allow them to impress their boss, as well as do a little bit of "networking". These may seem ephemeral needs, but studies of offices show that such practices oil the wheels of organisational process (Harper, 1998). It is not easy to see how such affordances may play a role in home settings.

6.1.2 An Approach to Paper-mail

Irrespective of whether there are complete parallels or not, these studies of office life suggest how one might look at the properties of paper-mail with regard to home settings. The research that generated these insights into the role of paper in office life required qualitative and observational research methods, which had hitherto not been utilised by those interested in the role of paper. In particular, a mix of ethnographic investigations, combined with a concern for the "interactional properties" of artifacts (which happened to be paper but could be any relevant artifact including computational) led to insights about the forms of interaction that people in office environments require (Harper, 1998; 2000a, b).

It was in light of this research that the Royal Mail funded a research programme at the University of Surrey's Digital World Research Centre (DWRC) which utilised the same qualitative approach to investigate whether there are similar interactional properties of paper-mail (similarly conceived of as affordances) which are and which may continue to result in the use of paper-mail in domestic environments. Some of the findings of this project, with a selection of materials gathered in simultaneous DWRC projects into smart homes, are presented here.

6.1.3 The Method

The following programme of activities, to be entitled *The Affordances of Paper-mail*, were undertaken. First, we undertook an ethnographic study of a panel of 11 households: 2 single households, 2 young couples, 2 older couples, a student household, and 4 families with children. The income ranged across the spectrum. Needless to say, though an attempt was made to ensure that a wide range of households were covered, given the total number that it was possible to look at in ethnographic work, it was not possible to obtain a truly representative example of all UK households.

The studies were undertaken over two periods, with the first being a pilot investigation and the second a more in-depth examination of what letters people chose to read, how the letters were moved around the home and why, and subsequent communications resulting from the opening of mail. Key to this analysis was a focus on the interactional properties of paper-mail, namely its affordances.

Data from this activity formed the basis of two other subordinate strands of research activity. The first of these was a small experiment. The experiment investigated some of the properties of searching and cataloguing envelopes, and this provided further insight into what the affordances of paper-mail might be by contrasting those provided by e-mail alternatives. The experiment investigated these questions in relation to the task of receiving and sorting mail, whether it be delivered in paper or digital media.

The second task was a small survey of about 200 persons. The questions used in the survey were put together on the basis of the ethnography and were designed to provide quantitative indicators about the frequency with which letters are used to support various patterns of social behaviour within homes, patterns which relied in one way or another on the affordances of paper-mail.

6.2 An Overview of Findings

In brief, the results of the research show that paper-mail does offer specific affordances that add value at the point of use over and above the affordances of other communications media, particularly e-mail tools as currently designed. Some of these affordances are ones important in office settings; others quite new. Perhaps of most interest, however, is the fact that some important affordances are those that support how members of households do things together. These may be called "social affordances".

For example, it has been well known for some time that certain types of mail are "broadcast" in the home. Postcards are an obvious example of this. Various attempts have been made to offer similar broadcasting of images in computationally mediated ways, e-cards being the least interesting. More creative ideas can be found in, for example, Liechti and Ichikawa (2000). The HomeNet project is also reporting some of the ways families "share" (Kraut et al., 1997). But this research showed that all types of mail can be shared within households. This was found in ethnographic data and the small scale survey which showed that women will share up to 57 per cent of the letters addressed to them (this includes all types of letters from personal to direct mail), while men an astonishing 69 per cent, including personal letters. Table 6.1 presents details of this.

The interesting issue here is not that they are broadcast, however; it is why. In summary, the reasons have to do with how letters in paper form are broadcast and moved around the house in a fashion that supports the social organisation of the family. Sharing or broadcasting letters is one element in this social organisation.

An interesting example – and indeed an unexpected use of sharing – is the way it is used by parents to monitor and control their kids. This monitoring can take surprising forms. Parents will not only sift out what they believe their children should or should not receive; sometimes they will ensure that their children know that this is being done. In our ethnographic data, one parent wanted to give a direct mail offer of a loan to her son so that "He would learn to throw it away". The affordance in question here does not simply consist of an ability to share; this affordance may be thought of as akin to the affordance of paper documents to oil the wheels of organisational life that was previously mentioned. This would seem an unlikely requirement for home settings. But what we found in our empirical studies is that such oiling of the wheels – in this case the wheels of family life – does indeed need to be done. Here it allows such things as parents to teach abilities and skills to offspring. Such didactic practices, sometimes resisted and resented no doubt, constitutes a key need (or function) of families.

There is a related affordance that our ethnography also uncovered and this has to do with how paper-mail has to be "bumped into". To illustrate: in one household we studied, the parents would open the teenager's mobile phone direct debit statement but knew that unlike more responsible members of the household, the teenager would not notice a statement judiciously placed on the kitchen table. Moreover, the teenager's asocial hours meant that there was little likelihood that the parents would be able to have a "handing over" moment when they could raise the question of who was going to pay for it. But the paper statement could be placed in

Table 6.1. Use of letters

(a) Do you ever show letters or other mail to other members of the family or make sure they see them?

	Show letters/other mail?		Total sample
	Yes (%)	No (%)	
Male	55 (69)	25 (31)	80
Female	74 (57)	55 (43)	129
Married/w. partner	99 (77)	29 (23)	128
Wid/div/sep/single	30 (37)	51 (63)	81
SEG ABC1	68 (67)	33 (33)	101
SEG C2DE	61 (56)	47 (44)	108
Children under 18	69 (76)	21 (23)	90
No children	60 (50)	59 (50)	119
All	129 (62)	80 (38)	209

(b) Show to other members of the family: what sort of letters or other mail do you tend to share with other members of the family?

Multi-coded

	Sample	As a percentage of	
		"Sharers"	Total sample
Personal/family/letters or cards	113	88	54
Business mail (tax etc.)	52	40	24
Catalogues	14	11	7
Business mail (for work at home)	8	6	4
Direct mail	4	3	2
Other*	16	12	8
Total "sharers"	129	100	62
Non-sharers	80		38
Total	209		100

* Bills, post cards, bank statements/receipts, "don't have any secrets", birthday cards, "depends on what I want them to see".

front of the teenager's bedroom door – so this is what they did. Now although the teenager could still manage to walk over the statement – after all it is not that great an obstacle – he could not do so without seeing it. And this meant that he was thereafter accountable for it. Either of the parents could then ask, "Well, what about that phone bill? Have you got enough money to pay for it?"; "What did I say about the price of the mobile phone?", and so forth. In these ways, then, the fact that paper-mail could be placed anywhere provided a key tool in the management of parent-teenager relations.

6.2.1 Managing the Home

The monitoring of kids is one thing, but using mail to monitor other members of a household often has to be more discreet than this. In another example from our ethnographic corpus, a wife monitored whether her husband had opened a direct mail catalogue that she thought might be of interest to him. Having identified the catalogue as of interest at the doormat, she then placed it where he would see it and then waited two days to see if he did anything. After two days, he had not done so, so she threw it away. In another family, the fact that after two days a husband had not done anything with a bill placed by his bedside prompted his wife to take up the task for herself. There were several other examples of similar practices of women managing men. Interestingly, such practices were picked up in the ethnographic research but remained less visible in our survey work. It might be that women are less than willing to declare their power in the home when asked to do so in a public place (the survey was undertaken in the street).

Irrespective of the problems of discovering these activities, what the examples do show is that the use of paper-mail turns out to be more like workflow control than in the earlier examples where mail was used to support family monitoring (where issues of discipline and learning showed themselves). Workflow is a grand term for technologies (typically electronic and interactive but not always) used to manage, co-ordinate and monitor tasks. Our findings show that paper can be one such technology in a household. Putting a bill on the kitchen noticeboard so that it gets noticed and paid may be thought of as workflow management, as is putting a bill inside a handbag so that it is found when one goes to the shops.

It is the corporeality of paper-mail that supports these "workflow affordances". Placing a bill in a particular place notifies all concerned what stage a set of tasks has reached. By the same token, the ease with which paper can be moved between points in the domestic workflow regime makes it a technology that can be used with minimal effort.

6.2.2 E-mail in the Home

The affordances of paper-mail that are relevant here might seem rather mundane. The fact that a letter can be seen to be in one place rather than another hardly seems a discovery worthy of the name; the fact that a letter can be moved easily is hardly a world shattering finding. But these properties do start to show their value if one compares them with what one can do with electronic alternatives.

Consider this: e-mail messages can be delivered to one person and presented on a single screen anywhere in the home. Now disregarding questions about what a message says, what our research suggests is that as soon as mail is sorted, recipients within households often start broadcasting it – or at least sharing it in one way or another. It is at this point that some of the differences between e-mail and paper-mail start to show themselves. Sharing may be supported in a sequential process with e-mail, when for example, a mother and child take turns with a screen. Alternatively, email can be shared concurrently, with various members of the household having their own screens in various places.

Yet either scenario has problems. In one of the examples above we saw that sometimes it is the physical handing over of a letter that is a key moment in the process of sharing or broadcasting in domestic settings. E-mail tools cannot readily support this: that is, though they can be used to send or forward messages, what they don't do is support the physical and ceremonial handing over so important in face-to-face situations. One can imagine how new versions of e-mail tools might do this. For example, many hand-held devices or personal digital assistants (PDAs) support the use of infrared signalling and the exchange of data between terminals. Such data could consist of e-mail. Thus one could image a mother summoning a child and beaming, while face-to-face, the "offending e-mail".

An issue here, though, is not whether this is possible to design – it certainly is, especially with the arrival of Internet-enabled hand-held communications devices. It is rather that currently this process of family monitoring is supported by the fact that the mother gets line of sight of all the mail that comes through the door. Thereby they can act as gatekeepers. If one was to offer e-mail to individual terminals and PDAs, this would no longer happen. And thus the mother would not know when their son's or daughter's direct debit statement has arrived (or indeed any other form of communication). As it happens this is one reason why teenagers are so keen on texting on mobile phones: it is because Mum (and Dad) cannot see what they are up to.

Texting aside, we have also remarked that where a letter is in the geography of the home is a marker of what point a job-to-do has reached. E-mail might support this if the screens are located in places that equate to locations within the domestic workflow. Unfortunately there are at least two reasons why this might be difficult to achieve. First, there would need to be screens in a host of places, and this may create economic difficulties on cost alone. But perhaps a more salient difficulty relates to how these locations are rather flexible and differentially graded. Sometimes the fact that a letter is in the living room means it is a job-to-do-today but at other times it simply allows a recipient to pick up a letter when they are, let us say, having a cup of tea. In other words, the same place can be used for more than one task. According to this view, for e-mail to offer an equivalent affordance to paper-mail, not only would there need to be many screens throughout the house, but users would have to forward messages to each screen dependent upon a complex of factors, some of which are ambiguous as the last example shows. In any event, our studies of smart homes have shown that screens tend to be used for entertainment services, rather than for "boring things". In other words, bills would have to compete with *Top of the Pops* for screen collatoral. When bills are in paper form no such competition exists.

There is another issue over and above the allocation of messages to the right screen. This has to do with whether e-mail can allow members of shared households to monitor one another. The physical demonstrability of paper-mail results in what one might call "system state monitoring" being done unobtrusively and easily: a wife can see at a glance that her husband has not done anything with a bill by the bedside, for example. With screens, such monitoring would become more difficult and intrusive: a wife would have to look over a husband's shoulder when he or she is doing his e-mail, for instance, and what then for the delicate balance of power – and more importantly symbolic power – within a marriage?

A key property of paper-mail is then that it acts as a successful technology because it *fits* into the physical organisation of the home easily. E-mail alternatives could deliver mail but would not necessarily provide the embodiment that facilitates the intersecting of space and social roles within the household. Attempts to improve e-mail alternatives through offering numerous screens for viewing would provide some benefits, though would create new screen collateral constraints. Whatever their design, it is probable that they would still not replicate all the affordances of paper-mail. In any case, there would be some added burden in terms of screen navigation techniques, and in terms of how one user would be able to monitor another user. Mobile devices would create new problems, both in terms of obviating opportunities for monitoring and allowing for the dispersal of e-mail messages that might undermine some of the workflow management tasks necessary in the home.

6.2.3 Changing Users

Although there might be difficulties, there could be ways of forcing e-mail in to the home which would not focus solely on the problems of designing technology. Three come to mind.

First, recipients could change their mail-related behaviour. Currently, they subject their mail to what one might call a process of triage which involves somewhat casually planning out some things "to do now" and some "at a later time", and then following on from this, using the affordances of paper to support the domestic workflow. Instead, recipients of e-mail could be more instantaneous in their reactions, paying bills as soon as they arrive for example, and managing the workflow within the home in a rather heavy-handed way: a wife would not simply watch to see if her husband does pay a bill, say, but would pointedly monitor his e-mail in-tray in ways we have mentioned.

As it happens, one can imagine many utility organisations being very pleased if the response of consumers to the arrival of their bills was more prompt than it is now. One study by Pitney Bowes (2000) has suggested this will be the case. Here, a pilot group of customers did indeed make payments more quickly than before when given EBPP. Unfortunately, that they did so is precisely what one would expect if for no other reason than the so-called "Hawthorne Effect". According to this, subjects will alter their behaviour simply because they are being watched (see also Rubens, 2000, for a more subtle review of the issues related to e-billing). Whether these same subjects will continue to behave in the same way is, we think, quite doubtful. Once over the initial interest in the new method of payment, it is unlikely that recipients of mail (i.e. the consumers) would accept this. Though both e-mail and paper-mail could be technologies that help sustain the business and social affairs of the home, paper does so in a way that allows members of the home to remain in charge. With paper, members of households can do things when they want at the speed they want; paper also allows them to monitor this without being intrusive. In contrast, e-mail could force members of households to behave in accordance with the wishes of the letter's sender: something that does not fit into what is sometimes called the "natural order of the home" (Rouncefield et al., 2000; O'Brien and Rodden, 1997; O'Brien et al., 1999; Hughes et al., undated). Given these disadvantages, cost incentives could be provided to encourage these changes in behaviour.

A second scenario involving some change on the part of recipients seems much more likely. Here, the future of e-mail in the home is one where users simply convert e-mail into paper-mail. So if bills, statements, mandates, certificates and other communications were to be delivered electronically, recipients would choose to print them because it is in the paper form that they can be moved around, handed over, cross-referenced and left in certain places to ensure that what needs to be done gets done.

If this turned out to be the case, then it would have a host of implications for the ways in which some of the things conveyed by mail would be supported in the future. Consider branding. If it were the case that recipients printed their mail, then differentiating a brand through, let us say, quality of paper and printed images (such as logos) would be obviated, since all the mail will be printed on the same device on standard grades of paper.

In any case, a more likely consequence of this scenario is that users would eventually tire of the hassle and cost of printing mail for themselves and would instead return to the practice of waiting for mail to arrive through the door. But they would also want to receive e-mail too. The reason why domestic users would want e-mail as well as paper alternatives is that the arrival of e-mail might help facilitate the delicate management of their domestic responsibilities: if in the past they would plan their response to mail through reference to, say, the colour of a bill – blue for "Put aside" and red for "Do something about it" – with e-mail they might be able to create a third level of reminder: "Oops! they are sending e-mail now, I really better pay", one can hear them say. This would be an especial problem for billing organisations since it may well be that users will also opt to continue having e-mail versions of the bills as well as traditional paper versions. Thereby the total costs of the sending organisation will increase rather than decrease.

However, a third scenario does come to mind, which is perhaps the most radical of all. This would involve users keeping the best element of current paper-based communications alongside what new technologies might offer, though the success of this would depend very heavily of the businesses being more subtle about how they use business to consumer (B2C) communications than would appear to be the case at the moment. For example, one might imagine a scenario where an organisation using paper-mail recognises that some types of information or product offering should be sent to the household, rather than to some particular person within a household. One can imagine what sort of products and information this may consist of, broadly speaking related to the running of the home, and maintaining domestic workflow. Businesses would have to be careful when they send such mail, however, particularly when it comes to addressing, since they need to recognise that most probably they are sending correspondence to the wife-girlfriend-mother when they send to a household, rather than say any one else in the home – such as the husband. Women may not want their actual status and power made too explicit in communications. As we mentioned, homes contain delicate balances of symbolic and actual power. A misaddressed letter may upset this.

At the same time, one can also imagine other communications and product offerings being sent directly to individuals PDAs and mobile devices. These sorts of products may have nothing to do with the home and all the tasks related to it and more to do with matters related to that person's "away from the home" world. We have not explored what this might be here, but what should be clear is that the disjunction between these two worlds is not at all what it might seem and is related to family structure, age, gender and much else beside. Consider the example mentioned above where we saw how the use of mobile phones by teenagers outside of the home is a matter of concern for those who run the home within, that is, Mum and Dad. This simple example attests to the complexities of the divide between home and elsewhere. All the more necessary, therefore, that those wanting to leverage new opportunities with current and future technologies take a more research-based approach to defining how they might explore new B2C opportunities, whether it be via paper-mail or digital alternatives, including mobile ones.

6.3 Conclusions

Needless to say, what we have done here is used empirical findings to ask more questions than we had started with: if one of the appeals of the substitution argument mentioned at the outset is its elegance analytically, then one of the problems with the qualitative methods we have been using – primarily but not exclusively, ethnography – is that they don't always allow us to come up with easy answers. (For discussion of the general arguments here see Bannon 2000, pp. 230–40). But, more importantly, this research was done in the UK and there may be cultural factors shaping the use of mail – and hence the affordances of relevance – that are different in different countries. For example, the number of bills received by households in the UK is small compared with the USA. This may change the behaviour of people when they receive bills so it could be dangerous to assume that reactions in the UK will automatically track into the USA. However, many of the aspects of the research, such as women managing the home, do seem familiar to Americans.

These concerns notwithstanding, in this chapter we believe we have presented some findings that may make those who use mail in whatever form think about what actually happens when mail is received, as well as offered some insights for those who are in the business of designing new communications media, particularly media that might support what has come to be called smart home technologies. We have pointed towards the problem of how domestic e-mail tools need to support workflow, for example, and how current e-mail tools simply don't get designed with workflow issues in mind, except insofar as they allow a serial distribution of activities. Workplaces might be organised in something like that fashion, but homes surely are not. As is well known, one of the problems of workflow technologies is related to the question of corporeality, or the lack of it. That is to say that when workflow "objects" are limited to being virtual, then some of the social organisational properties of distributed tasks are rendered opaque to participants in those tasks. This is one of the reasons why digital-only workflow tools nearly always fail (Abbot and Sarin's 1994 paper being the classic explanatory text of this problem). Yet in the home, the need for corporeality – and all the associated affordances which go with it – is as much a ceremonial requirement as it is a prerequisite for members of the family to be able to monitor just whose job is whose. Of course, members of families don't want to burden themselves with a frame of mind that they may adopt at work and which says, "I need to take account of my responsibilities". One doesn't go home to take on a new job, after all. But in practice home life is indeed just like work: it is socially organised and people do rely on each other in often complex and subtle ways to share and distribute tasks. But these are the tasks of family life and family living. Home life requires working at too though this is not at all like the work one does at work. That distinction notwithstanding, in this chapter we have sketched out what some of the characteristics of that social organisation of the mail-related work in the home might be.

References

Abbot, K and Sarrin, S (1994) "Experiences with Workflow Management: Issues for the Next Generation", *Proceedings of CSCW '94, Chapel Hill, NC*, pp. 113–20, New York: ACM Press.

Bannon, L (2000) "Situating Workplace Studies Within the Human-computer Interaction Field", in P Luff, J Hindmarsh and C Heath (eds.), *Workplace Studies*, Cambridge: Cambridge University Press, pp. 230–41.

Coopers & Lybrand for the European Commission (1996). *Workshop on the Impact of Electronic Mail on Postal Services*. Brussels: European Union.

Coopers & Lybrand Consulting (1998) *Substitution Trends and the Market Potential for Communication Distribution Services*. Canada: Canada Post Corporation (4 March).

De Rycker, T (1987) "Turns at Writing: The Organisation of Correspondence" in J Verschueren and M Bertucelli-Papi (eds.), *The Pragmatic Perspective: Selected Papers from the 1985 International Pragmatics Conference*, Amsterdam: John Benjamins Publishing Co.

Frohlich, D, Chilton, K and Drew, P (1997) "Remote Homeplace Communications: What Is it Like and How Can We Support It?", in H Thimbleby, B O'Conal and P Thomas (eds.), *Proceedings of People and Computers 12*, pp. 38–41, Godalming: Springer.

Harper, RHR (1998) *Inside the IMF: An Ethnography of Documents, Technology and Organisational Action*, London and San Diego: Academic Press.

Harper, RHR (2000a) "Getting to Grips with Information; Using Ethnographic Case Materials to Aid the Design of Document Technologies", *Information Design Journal*, Vol. 9, No. 2–3, pp. 195–206.

Harper, RHR (2000b) "The Organisation in Ethnography: A Discussion of Ethnographic Fieldwork Programs in CSCW", *CSCW: An International Journal*, Vol. 9, pp. 239–64.

Hughes, JA, O'Brien, J Rodden, T, Rouncefield, M and Viller, S (undated) *Patterns of Home Life*, Department of Sociology, Lancaster University.

Izutsu, I and Yamaura, I (1997) "Effect of Telecommunications on Letter Mail Services in Japan", proceedings of the *Conference on Postal and Delivery Economics*, 11–14 June, Helsingor, Denmark.

Kraut, B, Sherlis, N, Maning, J, Mudkophadhag,T, and Kiesler, S (1996) "HomeNet: A Field Trial of Residential Use of the Internet", *Proceedings of CHI 97*, pp. 77–90.

Liechti, O and Ichikawa, T (2000) "A Digital Photography Framework Enabling Affective Awareness in a Home Network", *Journal of Personal Technologies*, Special Issue on Domestic Computing, Vol. 4, No. 1, pp. 7–21.

Nikali, H (1998) "Replacement of Letter Mail by Electronic Communication to the Year 2010," in MA Crew and PR Kleindorfer (eds.), *Commercialization of Postal and Delivery Services: National and International Perspectives*, Boston: Kluwer Academic Publishers, pp. 223–35.

O'Brien, J and Rodden, T (1997) "Interactive Systems in Domestic Environments", in *Proceedings of DIS'97*, Amsterdam: ACM Press, pp. 247–59.

O'Brien, J, Rodden, T, Rouncefield, M and Hughes, J (1999) "At Home with the Technology: An Ethnographic Study of a Set-Top-Box Trial", *ACM Transactions in Computer-Human Interaction*, Vol. 6, No. 3, September, pp. 232–308.

Pitney Bowes (2000) *D3 Digital Document Delivery*, Powerpoint slide set, Pitney Bowes, London.

Plum, M (1997) "The Challenge of Electronic Competition: Empirical Analysis of Substitution Effects on the Demand for Letter Services", in MA Crew and PR Kleindorfer (eds.), *Managing Change in the Postal and Delivery Industries*, Boston: Kluwer Academic Publishers, pp. 133–61.

Rouncefield, M, Hughes, J, O'Brien, J and Rodden, T (2000) "Designing for the Home", in R Harper (ed.), *Personal Technologies: Special Issue on Home Life*, Amsterdam: Kluwer, pp. 76–94.

Rubens, S (2000) "Statement Design in the New World of the Internet", *Xplor Conference on Electronic Billing and Statements: Industry Trends and Practices*, London Heathrow Excelsior, 8 February.

Sellen, AJ and Harper, RHR (1997) "Paper as an Analytic Resource for the Design of New Technologies", *Proceedings of CHI '97 Atlanta, GA*, New York: ACM Press.

Sellen, AJ and Harper, RHR (2001) *The Myth of the Paperless Office*, Boston, MA: MIT Press.

Switching On to Switch Off

7

Alex Taylor and Richard Harper

7.1 Introduction

The burgeoning technologies that are emerging from the convergence of broadcasting, telecommunications and computing promise significant changes. Devices such as interactive TV and TV-on-demand promise to provide those in the home with unprecedented access to information and entertainment. They also raise the prospect of altering the relationships between home activities and those traditionally undertaken elsewhere – work, shopping and play.

Until now, however, relatively little has been published about television viewing and interactive TV in the system design literature. The disciplines that focus on the interaction between people and technology have been primarily oriented towards studying work and particularly office-based activities. Where domestic activities such as play and entertainment have been considered, solutions have generally been driven by technological advances rather than an understanding of the interactions between people and technology in the domestic context. Such an approach does not tend to consider why people use technologies such as TV and the relevance the technology has in people's everyday lives (Norman, 1999).

To counter the lack of in-depth social research into TV viewing in the home, a small number of studies have sought to use qualitative field studies to explore the relationships between technologies and people's daily lives (e.g., Black et al., 1994; Logan et al., 1995; Mateas et al., 1996). Despite their use of these explorative techniques, however, these studies have usually aimed to elicit user requirements for specific technologies rather than gain a general understanding of how TV plays a role in domestic life. Thus, although they provide a reasonable basis for designing usable television user interfaces, these studies fall short of explaining what it is about television watching itself that influences how viewers interact with televisions.

In light of the above shortcomings, the research we report here is specifically targeted at investigating what we call the natural rhythms of TV viewing. By this we refer to the common, ordered and patterned use of TV in the home – to the taken-for-granted practices of what has come to constitute TV viewing. This inquiry is oriented towards how an understanding of these practices might be used to inform design and specifically the design of systems for programme selection and storage.

7.1.1 Method

The presented research has drawn on three methods for collecting data: focus groups, household interviews and ethnographic fieldwork. These activities were undertaken in serial order, with the focus groups providing a basis for the household interviews, and both the focus groups and the household interviews providing a metric for determining what to seek in the in-depth ethnographic research.

In total, six focus groups were held in three different regions across the UK. The topics raised and discussed in the focus groups included typical evening viewing, programme choice, video use and future technologies. The household interviews comprised of visits to 20 families and investigated the ways people view TV, gain information about programming, navigate their way around their systems, select programmes, and use and store videos. Both the focus groups and interviews consisted of people from a range of age groups and socio-economic backgrounds. People were also selected based on their adoption of existing technologies including terrestrial TV, PCs/interactive technologies and multi-channel satellite/cable TV.

The aim of the ethnographic fieldwork was to provide rich qualitative descriptions of how people go about choosing programmes and watching television in the context of their own homes. Specifically, eight households took part in exercises to learn why and how people watch TV, and how TV is seen to fit into daily life and commonplace, domestic activities.

7.2 Findings

7.2.1 TV Viewing

From the results of the data collection, it was no surprise to discover that for the majority of people, the television was a near-permanent feature of their home lives. Indeed, across all social-economic groups it was seen as a near constant companion.

> In my household there is no difference . . . when I'm there, the TV is on. Even when I'm working in the house (and there's) a lot of background noise, I need it to be on . . . even when I'm in the kitchen cooking or washing up (Female, under 45).

This use of the TV, however, says very little about what television watching entails as a social activity. What it does indicate is that television is not viewed as something that is special or unique, so much as a natural and common feature of the home. This is important as it suggests that TV watching is not to be thought of as something akin to, for example, watching a video: that is to say, an event unto itself. Even though the TV is sometimes used precisely for that, the ubiquity of the TV watching makes it distinct. That is to say, the TV is "on" but not being watched; it is part of the furniture and treated as such. This has important consequences: The suggestion that the TV is part of the furniture is indicative of how its function in the home is prosaic and taken for granted in much the same way as a table or chair or a picture on a wall might be; a distinction of course is that the TV can sometimes deliver "content" that makes it more central than at other times. But rarely does this content engage or absorb members of the household's attention in the way that a video might, or say the Internet. This is not to claim that the TV is rarely attended to, it is to grade that level of attention.

This attention has certain particular characteristics, a "practical rhythm" of TV viewing is itself made up of pieces, or periods. From the earliest focus group interviews through to the ethnography, it was found that viewers tend to establish regular patterns of viewing. We found this to be especially the case on weekdays, during the late afternoon and evening. Daytime and weekend viewing were far less structured and were highly dependent on such things as weather and the season. Concentrating on this patterned weekday viewing, we found that most households had three distinct periods of television viewing: the "coming home" period; mid-evening viewing; and later-evening viewing.

7.2.2 Coming Home Viewing

Coming home viewing normally began after work or school in the afternoon or early evening. The TV was turned on to unwind, to start the process of relaxing or as a form of distraction, undertaken alongside other activities. For want of a maxim, this behaviour could be described as "switching-on-to-switch-off". Generally it can be characterised as highly disengaged viewing.

> As soon as I get in the TV is turned on, and we're not necessarily watching it but the TV is just turned on. We might be . . . on the phone to somebody, or our friends are round, but the TV's still on. I can't say there's a time when the TV's not on to be honest. That's it really (Male, West Midlands, under 45).

People were also very tolerant of what they watched during this period. For the most part, programmes were chosen in a highly unplanned fashion by "surfing" through the channels until something appealing was found.

> I will turn the television on and just flick through the channels when I come in, and probably keep it on and wander around doing whatever I've got to do (Female, west London, 45+).

The participants in our research claimed that their main method for selecting programmes during this period was to channel hop – switching or "surfing" between the channels searching for something that appeared interesting or familiar. The focus was on choosing something to watch now or possibly next (the latter achieved by catching programme previews or announcements). Notably, people made little to no use of programming guides.

7.2.3 Mid-Evening Viewing

The next period, the mid-evening viewing, would often run through dinner, and would last until about 8.30 to 9.00 p.m. In contrast to coming home viewing, this period had an order, with the planned viewing of certain programmes and with higher levels of engagement. During this period, household members chose programmes that they regularly watched, like soaps, sports, game shows or the news. Content providers call this "viewing by appointment".

These programmes would often be viewed communally and would also dictate when and where other household activities, such as dinner and homework, took place.

> Actually, if there is something very good [on], and I . . . want to watch it, I prepare dinner earlier so that we finish by the time the programme is on (Female, west London, 45+).

> I've got a through lounge so I always make sure that my dinner is prepared just before EastEnders comes on (Female, south-east London, 45+).

During the mid-evening period, where levels of engagement varied, viewers relied on their knowledge of the programming schedules to choose what to watch. Specifically, they relied on their daily or weekly routines to help them remember what was on. This habitual time-based selection generally involved viewers knowing that particular types of programmes were on at specific times. Occasionally, viewers would also make mental or physical notes of the programmes they wanted to watch, such as subsequent episodes of a documentary series or drama. Both these methods allowed them to turn directly to the desired channels without the need for programme guides. Only after regularly watched programmes had finished, did people during this period make use of programming selection methods like channel surfing or reading through programme guides.

7.2.4 Later-Evening Viewing

The third period, later-evening viewing, would often take place once the day-to-day chores in the house were completed and last until 11.00 or 11.30 p.m. For example, several parents who participated in the research project said they would only sit down in front of the television and think about what they wanted to watch after they had finished dinner and put the children to bed.

This viewing tended to involve a relatively high degree of engagement in most households. People seemed to have specific types of programmes they wanted to watch after this later-evening "watershed". Documentaries, current affairs programmes and dramas were particularly popular. It was evident that household members would often have their own individual preferences at this time of the evening. It is worth noting that we found it common for households to have several sets – on average an amazing 4.1 in a survey of 5,000 people we undertook – and that this tended to reduce or eliminate any arguments about what was watched.

During the later-evening viewing people participating in the research tended to use programme guides more often. Predominately, viewers would use paper-based guides; however, the use of onscreen guides occurred occasionally. The guides would primarily be used for short-term planning. To select a programme, people would glance across the guide, looking specifically at shows that were currently being shown or on next. As well as the guides, people also channel surfed, particularly when they did not have immediate access to a guide.

7.3 Analyses of the Three Types of Viewing

From this description of the three distinct viewing periods, it is apparent that people watch television in quite different ways. These are based on the degree of engagement and the extent to which viewing is planned. Levels of engagement vary between the three periods starting low, then becoming variable and peaking in the late evening.

Nonetheless, television viewing appears to be curiously "unplanned". Unplanned in the sense that though they might know what they are about to watch – say during the mid-evening viewing – they do not at any particular point settle down and plan that activity with reference to programme guides. There is nothing that one might call a rational decision-making process.

Crucially, across all three viewing periods people organised themselves to minimise the amount of effort needed to choose a channel. Those participating in the research indicated that they had "thresholds" delimiting the effort they were willing to make to find and select programmes. It was notable that these effort-thresholds varied depending on the contexts people were in. Viewers watching television in the early evening, for example, had relatively low thresholds because they were tired from work and simply wanted to relax and unwind. Later in the evening, they claimed they would be more critical about the programmes that were on and would be willing to exert more effort in choosing a programme. This variability on the effort made by TV viewers resulted in a range of programme selection methods, ranging from channel surfing to the use of the paper or electronic programme guides (EPG). The patterned and seemingly ordered use of these methods is discussed in detail below.

7.3.1 Programme Selection Methods

Throughout the data it was evident that viewers tended to use programme selection methods in a specific order (Figure 7.1). Viewers began their search for a programme by channel surfing. If they failed to find anything using this method they searched – or waited – for a programme announcement to find out what was on next. These two methods, whose combination we call "Now-and-Next" were massively predominant. If they failed, however, viewer's knowledge of the weekly schedules or of upcoming programmes would be used. After attempting these three methods, the viewer would turn to either the paper-based or the onscreen guides. This order was not strongly fixed, and occasionally viewers would find themselves in situations where one or more of the methods were not appropriate.

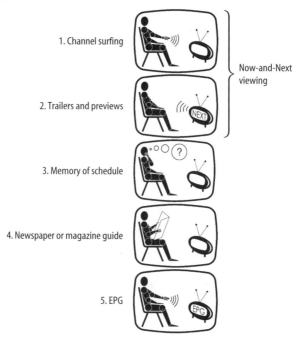

Figure 7.1 The sequence of methods used to make programme selection.

Nonetheless, there seems to be a certain logic to these methods, reflecting in part the social context of viewing – when, for example, they come home and switch the telly on to "switch off", as against switching on for viewing by appointment later on in the evening. At the same time these social contextual factors appear to be related to what one might call the cognitive load involved in using each type of method. For example, channel surfing was the first and most frequent method used because it was felt, by viewers, to be "effortless" and required little thought.

To understand why there was this perception of effortlessness, channel surfing must be considered in the larger context. From such a perspective, channel surfing can be seen as part of viewing. It is inherently associated with the act of "watching" television. When viewers turn the television on, they are immediately faced with a choice of channels and the act of watching necessarily involves navigating away from and thus toward a programme of choice. The navigation, in this sense is immediately "at-hand". Through this understanding, they recognise that by moving (or surfing) through the channels they will see what is on. It could be said that channel surfing is afforded in the act of watching television.

The other ways people select programmes require quite different interactional processes, each with increasing demands on the viewer. Although reading through a paper-based programme guide, for example, may not be taxing, it requires that the viewer step out of the act of watching television. In doing so, some of the affordances that were present in television watching are lost. Fortunately, reading and looking through information on paper is a familiar task for most people. Indeed, paper has been shown to have a number of properties that support reading and the navigation of information (Haas, 1996; O'Hara and Sellen, 1997). Consequently, reading through a paper-based guide itself is not demanding. Nevertheless, switching between the television and paper guide demands a transition in the way viewers think about choosing what to watch.

Switching to EPGs, such as Teletext and the OnDigital TV guide both of which are provided to UK audiencies, appears to require a more significant transition. This explains why people in this research were not frequent users of onscreen guides and tended to use the method last. Not only is a transition necessary with onscreen guides, but the understanding of the workings of the process are also unfamiliar. Furthermore, the operations can interfere with how viewers understand the television to work; the buttons on the remote control, for example, no longer work as expected. Studying the use of several EPGs, Daly-Jones and Carey (2000) have confirmed that viewers find EPGs difficult to operate. Specifically, they found that viewers often made mistakes when using the remote control to access programme information. They also discovered that viewers had difficulty in getting into and out of the information services.

7.4 Lessons for Design

The programme selection methods described above have several implications for the design of next generation programming guides. These
implications are discussed in the following sections.

7.4.1 Primary EPG

Our research into television viewing indicates that there is a common
process people use to choose programmes. This process tends to be used
in a set sequence that appears to be associated with people's perception
of the effort needed to step out of the act of television watching. It seems
that people choose information sources that require the minimum effort
to make the transition from viewing to choosing a programme. They do
this by using sources that are "at hand" and that make the decision-
making process simple.

This process raises several important implications for the design of
EPGs. Perhaps the most significant implication for EPG design is that
people have a preference for information sources that do not distract
from the act of watching television. This suggests that an EPG will only
be a viable solution if it can limit the disruption to people's sense of
what television watching is about. To do this the transition from viewing
to the EPG must not be perceived by viewers to be cognitively taxing.
One design requirement could thus be that EPGs make use of the same
perceptual modality people use to watch television. That is, the EPG
should display programme options not as text but as images maintaining
the visual-spatial modality. This could be achieved by displaying thumb-
nail images of the possible programme options.

Another way of reducing the cognitive demands associated with
using an EPG would be to simplify the decision making process. People
already do this by limiting the number of channels from which they
choose, to about five (though our evidence does not allow us to explain
this limit). They also predominately use the first two approaches to
channel selection: channel surfing first, and then trailers and previews.
The combination of these we have called "Now-and-Next" viewing. These
strategies could be supported through an EPG's interface. The
programmes that were on now-and-next could be displayed as thumb-
nails for a viewer's five favourite channels.

An example of an EPG interface incorporating these design suggestions is presented in the Primary Programme Guide in Figure 7.2. The
underlying idea to this design is that is provides viewers with quick and
easy access to the information they refer to most frequently. It is thus
referred to as the primary EPG.

It should be noted that this interface is only an initial indication of
how a design might actually operate. Specific usability tests would need
to be undertaken to evaluate any design suggestion derived from this

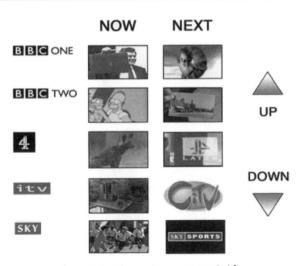

Figure 7.2 Primary Programme Guide.

exploratory research. For instance, further research would need to be done to determine how people should access the Now-and-Next EPG. Allowing viewers to switch to the EPG through a single button press on their remote controls, for example, would be in keeping with the aim to minimise effort. However, this design suggestion cannot be substantiated with the existing data.

7.4.2 Secondary EPG and Reviews/Editorials

Although people predominately use information sources to choose programmes that are on now and next, there are of course times when sources are needed for more detailed programme information. For example, viewers might want to find what is on later in the evening or may wish to get further information on something they are currently watching. They might also want to read editorial pieces or reviews associated with programmes they believe might be interesting. To provide a solution for this, a secondary EPG must be considered that augments the primary system described above.

Several findings from our research into television viewing can be used as a starting point for the design of this secondary EPG. People's comments about existing paper-based guides, for example, suggest that extensive programme listings should still facilitate quick and easy access to information. People liked the way they could glance at an entire day's programme schedule in the paper guides to find out what was on. The channel-time layout frequently used in these guides appeared to afford this "glanceability" because people needed only to interpret this familiar and easily understood technique for displaying information. If an EPG displayed programme information in a channel-time matrix, it too would presumably take advantage of people's ease of interpreting information displayed in this way. Of course problems arise with this design suggestion. The limitations of resolution and screen real estate, for instance, constrain the amount of information that can be displayed. Solutions designed to display channel-time matrices taking these constraints into account would need to be carefully evaluated before there could be any certainty of their success.

It is not so clear how an EPG could be designed to accommodate people's access to programme reviews and editorial. Magazine or newspaper guides are not considered by viewers to be extremely successful at displaying this type of information. People often have difficulty finding specific reviews or editorial using these guides because the organisation of the information is not "transparent". The techniques used to display the information are also not consistent between guides, making it difficult for people to establish familiar patterns of use. Paper-based guides thus provide few clues for how an EPG might enable access to reviews and editorial.

Another difficulty with designing EPGs for this purpose is that it seems people do not base their choice of programmes on the reviews and editorial they read. It may be that the reading of reviews or editorial materials is part of the separate guide-browsing activity. It is not entirely clear what people get out of this activity. Not having a full understanding of this makes it difficult to know how to design an EPG that meets people's needs in this context. It may be that part of the appeal of the activity is based on sitting back with the newspaper or a magazine and that an onscreen system could not provide the necessary affordances to be used in this way.

7.5 Conclusions

Numerous EPGs are available on the marketplace at the current time, bundled up with various set-top offerings. Our sponsor, who has some indirect commercial interest on the impact of these EPGs, wanted to know what might be the kind of design principles that good EPG design is based upon, and wished to test these against those used in practice. Our research shows not only that there might be cognitive loads that need to be borne in mind in EPG design, but also that these demands are related to the context of viewing; especially the three forms of viewing habits we have described. All current EPGs appear to be designed without reference to either the problem of cognitive load or this social context of use. Instead, they would appear to be designed on the basis of various rules of thumb developed on web-based information provision. This may well account for the low levels of regard that these EPGs are held in by the public at large, and indeed our sponsor's scepticism about them.

Of more importance, we believe, than the failures of the current crop of EPGs, is the approach to understanding user needs that we have presented. It is our view that good design should not only be based on the traditional techniques and concepts of cognitive psychology – such as notions of load and capacity – but should also take into account the kind of sociological materials that we have presented here, in this instance related to socially constructed habits and routines.

In addition to this interdisciplinary approach, we also believe that one should design for current practices in the first instance rather than for some posited notion of future user behaviour. In this case, although EPGs are expected to radically alter viewers' watching habits – especially when combined with local storage devices – it is our view that those changes are less likely to happen if the initial form of EPGs is so alien to current practice that users find them all but irrelevant to their current viewing habits. If EPGs were designed for how people currently behave, they could not only find acceptance but might also be designed to lead users towards new forms of viewing in a gradual way. When they first use EPGs, users can get familiar with their particular interaction modalities, they can learn what the guides afford in terms of new ways of navigating to programme choice, and so on. At a later date, new releases of EPGs can then move them further away from their original viewing habits toward new viewing patterns; these may be unlike the three-fold form we have described.

This might seem a pedantic way of designing for the future. It may be viewed as counter to the tradition of innovation and radicalness that pervades research in the digital technology domain in particular. In these settings, one often hears the phrase, "Users don't know what they want because they can't see the future". But in our research, we have found that taking users' current practices seriously has led us to uncover issues that can be of huge importance in ensuring that new services, products and technologies can be successfully introduced in the first place. We have focused here on home entertainment, but our research has also looked at many other areas too, as some of the other chapters in this book testify. We hope to have given some clue as to why this approach has enabled us to provide value and insight.

References

Black, A, Bayley, O, Burns, C, Kuuluvaineng, I, and Stoddard, J (1994) "Keeping Viewers in the Picture: Real-World Usability Procedures in the Development of a Television Control Interface", paper presented at CHI '94.

Daly-Jones, O, and Carey, R (2000) "Interactive TV: A New Interaction Paradigm", paper presented at the CHI 2000, The Hague, Netherlands.

Haas, C (1996) *Writing Technology: Studies on the Materiality of Literacy*. Mahwah, NJ: Lawrence Erlbaum Associates.

Logan, RJ, Augaitis, SR and Miller, RH (1995) "Living Room Culture: An Anthropological Study of television Usage Behaviors", paper presented at the Human Factors and Ergonomics Society 39th Annual Meeting.

Mateas, M, Salvador, T, Scholtz, J and Sorensen, D (1996) "Engineering Ethnography in the Home", paper presented at CHI '96, 13–18 April.

Norman, D (1999) *The Invisible Computer*, Boston, MA: MIT Press.

O'Hara, K and Sellen, A (1997) "A Comparison of Reading Paper and On-line Documents", paper presented at CHI '97, Atlanta, GA, 22–27 March.

The Social Context of Home Computing

8

David Frohlich and Robert Kraut

8.1 Introduction

Computer and Internet use in the home does not only depend on the functionality of available software and services. It also depends in a very practical way on how the computer itself is located, managed and shared between family members. These factors constitute the social context of home computing and form the subject of this chapter. We report the findings of a home interview survey with 35 families in Pittsburgh and Boston, in which family members spoke about the practicalities of using a computer and going online. The findings show a variety of ways in which the computer is being domesticated to fit into existing patterns of family life, home architecture and parental control. They also point to the significance of introducing a second computer into this situation, and its similarity to introducing a second television. The implications of these findings for the design of home technology is discussed.

8.1.1 Aims

Most discussions of domestic Internet use centre around the content and benefit of Internet *services*. Indeed, the prime objective of many recent research studies in this area has been to inform these discussions with data on the relative use and value of different services by a sample of families (e.g. Kraut et al., 1996). However, in the course of these studies it is becoming apparent that the way families use and benefit from the Internet is not simply a function of what they can do on it. These things are also influenced in a very practical way by the accessibility of the family PC as the primary means of "going online" today. For example, factors like who can get on the Internet, in which room, at what time and for how long in any family, are as important as what they can do on the Internet once they are connected. These factors relate to the social use of computers and time within the family, and have implications for the design of computing and Internet technology in the home.

In this chapter we examine this social context for home computing and its relationship to Internet use. After a review of other studies in this area, we introduce findings from two sets of in-home interviews with 24 Pittsburgh families and 11 Boston families. The Pittsburgh families formed about a quarter of the original families in the HomeNet trial of the Internet (Kraut et al., 1996) while the Boston families were part of an investigation of home PC futures within HP (Frohlich et al., 2001). The findings reveal a rich and complex set of behaviours with computing technology, which are aimed at *domesticating* it within existing patterns of family life.

8.1.2 Previous Types of Research

In contrast to the extensive literature on the social context of computer use in the workplace (Baecker, 1993), there is little written on the social context of computer use in the home. This is very much a sign of the times and a case of social science trying to catch up with changes in human behaviour resulting from rapid developments in technology. With hindsight we can now look back on the 1980s as an era in which the personal computer entered the workplace and began to modify working practices in fundamental ways – ways that we are only now beginning to appreciate and use in the development of better workplace technology. In the same way we will look back on the 1990s as heralding an era of home computing and Internet use with all its attendant influences on domestic practices and family life. Unfortunately we are far from understanding what these influences are today, and even farther from applying such understanding to the design of home computing products.

Inroads into this area have begun in a number of places and serve to set the context and questions for our current enquiry. Essentially they have been made in three areas relating to the use of time, the use of space and the use of technology in the home.

8.2 The Use of Domestic Time

A large number of studies dating back to at least the 1950s have investigated the use of time using time diaries (Robinson, 1988). Subjects in the studies are usually asked to fill in a diary of what they are doing, where and with whom every 15 minutes throughout the day, and these entries are then coded into 100 standardised activities. The activities cover things such as paid and unpaid work, caring for children, obtaining goods and services, sleeping, washing, dressing, eating, learning, organisational involvement, entertainment, recreation and communication (see

Harvey et al., 1984 for an explanation of methods). Studies are often large national or multinational time use surveys, comparing broad patterns of time use between different parts of the population. Furthermore, the same studies are often repeated at regular intervals, perhaps as part of a national census, so that time use trends can be monitored. In the context of this chapter, we are most interested in localised patterns of time use within American households. Robinson and Godbey (1997) provide the best account of this behaviour, although this is based mainly on the analysis of three national US surveys conducted in 1965, 1975 and 1985.

Most time diary studies, including those examined in Robinson and Godbey, show that human activities are organised into recurring patterns or routines. Sleep, personal maintenance, work and recreation (especially TV watching) dominate American adults' use of time. The structure imposed by biology and culture causes some similarity in the cycle of these activities between different people. Biological disposition affects rates of metabolism and energy levels over a 24-hour cycle. Most people sleep at night and are awake during the day. External institutions such as employers, school and church demand people's presence at particular times of day. As a result, people go to work and school during weekdays, but have more flexibility in spending their time during the weekends. Television networks differentiate their programming for weekday and weekends, and for days and nights, based on predictions of the available audience during these periods. As a result, if working adults watch television, they are especially likely to do it during the prime-time hours of 8.00–10.00 p.m. on weekdays. And so on.

In the face of these broad similarities in schedules across people, there exist large individual differences between people, based on differences in the institutions they are connected to, on personal preferences, and on the composition of the household itself. Households with young children are likely to operate on a different schedule than household with no children or with teenagers present. People set their clock radios at a certain time get up to drive the children to school or go to work. Children have to be home at certain times set by their parents to eat or sleep. Parents have to coordinate their activities with childcare helpers and agencies so that their children are always cared for. In general, both the regular and irregular use of time by individuals is constrained by the number of other individuals they must live, work and interact with. Little wonder that vacations are needed from time to time to break from routine and literally "get away from it all"!

It is against this backdrop of daily routines and constraints that new technology enters family life. Somewhere within or between these routines, people must find time to use it. Here Robinson and Godbey's calculations of available free time at home are instructive:

If we characterize sleep and necessary eating and grooming from the 168 hour week for the economically most active segment of 18–64 year-old people in America, what is left are roughly 100 hours a week to divide between work, family care, other personal care, and free-time activities. A little more than half of that 100 hours (53 hours) goes to paid work and family care, a number that is surprisingly close for men and women. Another 40 hours are given over to free-time activities, almost half of which are devoted to the media, most of it to television; again the gender differences are minimal. The remaining 7 hours go to other personal care activities, such as the socializing that often extends meal times, the relaxing bath, or the grooming that is more vanity than necessity. One could also add here playing with children or window shopping, now coded as family-care time (1997, p. 293).

All this implies that up to 6 hours of free time are potentially available each day for home computing and Internet use, although nearly half this time is now spent watching TV and the other half is shared between socialising, home communication, reading, hobbies, outdoor sports and recreation, adult education, religious or cultural activities (see Robinson and Godbey, 1997, p. 125, Figure 12). Furthermore, the distribution of free time across the day depends on daily routines, which may fragment it into small pieces. So within the available free time of any individual there will only be a finite number of opportunities each day to use the computer and go online, and those opportunities must be taken at the expense of time spent on other free-time activities.

Although Robinson and Godbey's book is based mainly on time diary data, they make an excursion into a 1995 telephone interview survey on home computer and media use, specifically to explore home computer adoption (Chapter 10). According to reported time use estimates in this survey (which are less accurate than time diary accounts), home computer owners reported an average of 40 minutes computer use a day, of which 8.6 minutes was said to be spent online. Computer use was inversely correlated with TV use, suggesting that users may be borrowing from time spent watching television to use the computer. A recent Forester study drew similar conclusions after asking 100 PC owners directly how much they use the computer and where they find the time. The average user reported spending just under an hour a day on it, mainly at the expense of TV watching (Bass et al., 1996). A recent study by Nie and Ebring (2000) also suggests strong substitution between computer and TV use.

Given the limitations of these findings, and the absence of data on child and teen time use, it would be instructive to try to identify when different members of a household use the home computer and Internet, and what other activities they seem to be sacrificing to do this. In addition, it would be interesting to know whether these periods of computer and Internet use are slotted unpredictably into the gaps between established daily routines, or whether they are themselves becoming a routinised part of family life.

8.3 The Use of Domestic Space

While there is no single research field for the study of domestic space, a number of disciplines throw light on its use from different perspectives. These include archaeology, social anthropology, sociology, social and environmental psychology and computer supported cooperative work (CSCW).

Archaeological studies of ancient dwellings show them to have designs which reflect the lifestyle and culture of the inhabitants. For example, it is common to find palaces and temples at the centre of walled cities with roads radiating out to gates at each of four compass points (Wheatley, 1971). These links are even more evident in anthropological studies of living cultures where architecture, attitudes and behaviour can be studied together. Typically the arrangement of houses and rooms in a house reflects the social status of groups and individuals (e.g. Levi-Strauss, 1963). Furthermore, changes in house design often reflect changes in culture. Modern American and European houses evolved from semi-public medieval structures with a large central hall for receiving and entertaining visitors, cooking, washing, eating and working (Fairclough, 1992). In the eighteenth century, the open hall began to be partitioned into smaller spaces off a central corridor, like houses off a street. These rooms were named and specialised by function, and arranged according to a series of organising principles such as front/back, clean/dirty, day/night, public/private, sacred/profane (Lawrence, 1987). Eventually, a withdrawing room or parlour for entertaining visitors came to be placed at the front of the house near the door, kitchen and private living room areas were placed at the back of the house, with bedrooms and bathrooms located upstairs These arrangements afforded more privacy to individual family members, and underpin the relatively recent structures of childhood and the nuclear family (Aries, 1962).

The same themes of domestic space affecting and reflecting cultural practices and values are also evident at an individual level. People select, design and furnish their houses to support a current range of behaviours and interests pursued within the house. They also design to reflect their personality, and to present a variety of facets or "faces" to outsiders (Goffman, 1959). Spaces and objects in the house therefore have a mixture of functional, symbolic and sentimental value, all working together to make the house into a home (Csikszentmihalyi and Rochberg-Halton, 1981). When behaviours and personalities change, domestic space and objects must be reorganised to accommodate new requirements. This leads to a situation where buildings tend to grow with their inhabitants (Brand, 1995). This phenomenon is particularly evident throughout the life stages of a typical family, who begin with modest requirements for space which increase as children are born and grow up. This often leads families to extend or move "up-market" to a bigger house, although Friedman (1998) has shown that this could be avoided by building more flexible housing. His development of "Grow Homes" in Montreal comprises town houses organised into three tiered cells. Each cell has a large open interior which can be flexibly partitioned with mobile walls and furniture. As families grow, they can rearrange interiors and lease new cells in the house.

One particularly important use of domestic and other kinds of space is for social interaction. In fact space can be seen as a medium for interaction in much the same way as the telephone and e-mail can. Like these other media, space exerts considerable influence over the kind of interaction that can take place through it. At the most basic level, Osmond (1957) has observed that some spaces are more conducive to interaction than others. Some *sociofugal spaces* like railway waiting rooms tend to keep people apart. Other *sociopetal spaces* like street cafes tend to bring people together. Osmond, who ran a large health and research centre in Saskatchewan, commissioned a psychologist called Sommer to examine this phenomenon in his institution. Sommer (1959) conducted 50 observational sessions of conversations held around rectangular tables (36 ins. (72 ins.) in the cafeteria, noting who spoke most to who across the six possible seating positions. He found that corner situations with people at right angles to each other produced six times as many conversations as face-to-face situations, and twice as many as between people sitting side-by-side. Osmond and Sommer applied these findings to the arrangement of furniture in the hospital wards and dayrooms, by moving in small square tables to provide a place for reading materials, and maximise corner conversation. This resulted in twice as many conversations overall and three times as much reading by patients, with associated improvements in well-being.

As a side effect of Osmond and Sommer's intervention they encountered great resistance by patients to the movement or removal of "personal" chairs. This illustrates another feature of the use of space for social interaction: *territoriality*. Like other animals, humans have a tendency to take ownership of spaces and defend them from others. This was vividly demonstrated in another study of the use of chairs in old people's homes in South Wales. Lipman (1967) logged the proportion of time that dayroom chairs were occupied by their "owners" as opposed to others using the room. Chairs in regular use were found to be occupied by their owners an average of 93 per cent of the time. Occupants of the home actively chose to remain in familiar chairs despite opportunities to move to more comfortable positions out of the sun or in better view of the TV, and sanctioned others who moved into their chairs. This kind of territoriality also extends to the distance people keep between themselves and others. Hediger (1955) coined the term *personal distance* to refer to the invisible bubble of space people maintain around themselves in interaction. He calculated this distance at between 1.5 and 4 feet, which would place the other person within reach or at (2) arms' lengths away. Hall (1966) has subsequently expanded the concept of personal distance to include four distance bands, including *intimate distance* (contact to 1.5 ft.), *personal distance* (1.5–4 ft.), *social distance* (4–12 ft.), and *public distance* (12–25 ft.). Although the social significance of this classification is unclear, Hall is right to observe that as distance between people increases, basic changes in speech, hearing, gesture and vision take place which may affect the tone and character of their interaction in complex ways. In a more modern context, Heath and Luff (1992) confirm this in their studies of videoconferencing tools which effectively reduce the size of someone's perceived face and body on a TV screen. The character of conversation is subtly affected by lack of visible feedback from facial expressions, and regular users of the equipment learn to exaggerate expressions and gestures to compensate.

Finally, Heath (1986) has also shown that the character of social interaction is dramatically affected by the presence of computers. In several studies of doctor patient interaction he found that the introduction and placement of a PC monitor on the doctor's desk led the doctor and patient to behave quite differently towards each other. If the monitor was angled towards the doctor and away from the patient the doctor tended to orient his or her attention towards the screen at the expense of the patient. If the monitor was positioned so that both parties could see the screen, the doctor and patient could coordinate their attention to the screen and each other more effectively. These kinds of effects are now the subject of a number of studies to understand the role of physical artefacts of all kinds in social interaction, including paper, whiteboards, displays and furniture (e.g. Luff et al., 2000).

All these studies begin to show that finding space in the home to operate a computer and go online is likely to be a complex matter for any family. Not only must its location fit in with cultural and family norms regarding the use of different rooms in the house, its appearance and image must be consistent with the decor of the room and the personality of its users. Furthermore, on a more practical level, putting the computer in a more private space will give the owner of that space privileged user status, and discourage others from sharing the device and talking to the user. Likewise, placing it in a more public area will encourage greater sharing and interaction around the device, especially if the orientation of the monitor allows others to draw close enough to read text on the screen. This in turn may lead to lack of privacy for individuals, and contention for use.

Given the lack of data on these topics it would be interesting to explore where exactly families choose to locate computers for Internet access in the home, how they come to these decisions, and what experiences they report with operating the computer in different locations. Because of the concern raised in earlier parts of the HomeNet project with Internet use leading to increased social isolation, it might also be productive to explore the reported effect of computer placement on patterns of social interaction within the family.

8.4 The Use of Domestic Technology

A great deal of technology fills the home of the average American family. Washing machines, fridges, telephones and televisions are all-pervasive today – noticed more by their absence than their presence (Birnbaum, 1997). The same is not yet true of the computer which is still missing from over half the households in the USA, and remains a mystery to many. Birnbaum argues that the computer will ultimately be domesticated in the same way that electric motors have been domesticated; as a component of numerous home appliances which help people to do a well-defined task very simply. In his view, the general purpose home computer with optional Internet access will give way to a variety of focused-function Internet appliances, which derive their functionality from "information utility" companies that dispense software and content in the same way that power utility companies now dispense electricity or gas. An alternative view is that as PC prices continue to fall, more households will buy more attractive home computers. Given the current importance of this debate for technology providers and ordinary citizens alike, it is surprising that so little is known about how previous information technologies became pervasive and whether the home PC and the Internet are moving along the same trajectory. What clues there are come from research on the telephone, the television and a handful of studies on home PC use.

A number of historical accounts of telephone adoption stress the fact that the device came to be used in ways the inventors never imagined. For example, Bell's early demonstrations of his invention involved the relay of live musical performances from one place to another, without any dialogue in the opposite direction (Aronsen 1977). This radio model of telephone use was subsequently incorporated into a more suitable broadcasting technology, while the telephone itself became used for two-way conversation. Even here, the social value of telephone use was underestimated by service providers and consumer groups alike. Phone users were initially trained to use the phone as efficiently as possible for business transactions, and idle chatting was actively discouraged. Domestic use of the telephone for small talk was a later use which emerged despite rather than because of the promotions of telephone companies. Other aspects of these promotions stressed utopian notions of the telephone abolishing the effects of distance and removing class and gender stereotypes. In practice, the effect of the telephone, while massive, has tended to be less revolutionary than this, largely replacing the practice of letter writing for keeping in touch with distant relatives and friends, but not removing the need for local contacts or for face-to-face meetings (c.f. Welman and Tindall, 1993). As for gender stereotypes, the telephone appears to bring them into sharp relief; with women using the phone as a recreational tool for chatting and socialising and men using it as a tool for work and making social arrangements (e.g. Lacohee and Anderson, 2001).

TV use has been more extensively researched. Gunter and Svennevig (1987) draw together many of the findings from a variety of studies of using set meters, viewing diaries, interviews and video observation. TV adoption has appeared to move through three stages, where TV watching starts out as a community activity because of the scarcity of sets. As sets become more affordable, viewing becomes a family activity in the home, until prices fall so far that multiple sets can be purchased for the same household. Additional sets tend to be placed in adult or child bedrooms turning TV-watching into a more solitary activity, although adult-adult and child-child viewing remains prevalent (Bower, 1973; IBA, 1987). Both parents and children in the USA and UK tend to watch about 3 hours of TV a day, but viewing different programmes at different times (Ehrenberg, 1986). However, this figure disguises the fact that about an hour of this time is spent doing other activities concurrently. These activities include talking, eating, sleeping, reading and exercising (Betchel et al., 1972). Thus the TV moves from being the centre of attention for all the family at routine times throughout the week, to a background noise which exerts little influence on surrounding activity (Lull, 1980). In between, the TV can be a source of conflict and contention if family members cannot agree about what to watch next, or if parents and children disagree over the timing and suitability of certain programmes. In these cases it has been found that fathers tend to act as final arbitrators of viewing decisions, but will often defer to the wishes of their children (Bower, 1973; Lull, 1982).

PC use, on the other hand, has tended to evolve from a more solitary and specialised status in the home. Interviews and observations in the early 1990s conducted with 20 families in the south-east of England showed that their computers, if they had one, were used either for work or game-playing by just one or two individuals in the family (Silverstone, 1991). Alternatively they had fallen into disuse for want of appropriate expertise and interest. This situation has been changing rapidly in recent years with the increased penetration of computers into the home, the explosion of available software, and the advent of the Internet. Venkatesh (1996) is one of the few researchers to have tracked these changes in home PC use in America, through large-scale telephone surveys and in-home interviews. He claims that whereas home computers in the 1980s were used primarily for word processing, telework and children's games, home computers in the 1990s were being used for a wide number of household functions such as child and adult education, family communication, family recreation and travel, shopping and domestic finances. Furthermore, more members of the family are now engaged with computer use. Many of these findings are played out in detail in the HomeNet study itself, which shows widespread use of Internet services by each member of the family.

In exploring home PC use further, Mateas et al. (1996) show that many of the household activities now supported by the PC are normally distributed throughout the house in time and space, and may be carried out jointly rather than individually. Having to go to a single location, one at a time, to perform these activities, constrains the value of the computer and its ultimate domestication into family life. This leads them to recommend the fragmentation of the PC into a network of home appliances:

> ubiquitous computing in the form of small, integrated computational appliances supporting multiple collocated users throughout the home, is a more appropriate domestic technology than the monolithic PC (Mateas et al., 1996, p. 284).

Similar sentiments are echoed by O'Brien and colleagues from a series of home visits to ten PC-owning families in the north-west of England. They observed an "overloading" of the space occupied by the computer with activities normally distributed around the house, leading to competition for access and control. This led them to recommend distributed or portable computing technology for the home (O'Brien and Rodden 1997).

All this suggests a number of questions for the current analysis. The issue of overloaded space is important to understand further, since it appears central to the domestication of the computer in the home. In particular, we might ask how do families regulate conflicts for use of the PC and Internet when they arise? It is also interesting to note in this connection that PC adoption may be going the same way as TV adoption where households are beginning to bring additional PCs into the home (keeping older models) to meet increasing demand for use. We wonder how these second PCs are being used, whether they solve the overloaded space problem, and which PC is used for Internet access? If two is not enough, will the further domestication of the PC involve one for each member of the family?

8.5 Methods

To address some of the questions raised by previous research, we have combined the comments from two distinct home interview surveys. The first set of interviews was carried out in the homes of 24 families in Pittsburgh Pennsylvania between 1996 and 1998. These interviews were part of the HomeNet trial, which was designed to examine how a sample of households were integrating the Internet into their lives, during a period when the Internet was first moving out of research laboratories and academia and being used by the general public. Families were given or loaned a Macintosh computer, given instructions on how to use electronic mail and the World Wide Web, and were given a free telephone line and Internet access (see Kraut et al., 1996, for further details of the trial methodology). At least two researchers interviewed each of the HomeNet families to provide more qualitative information about use of the Internet to compliment the quantitative data collected through questionnaires and by logging Macintosh and Internet use. In particular, the visit interview schedule was designed to probe for typical patterns of Internet use in each household and provide opportunities for participants to tell stories of when and why they went online. Interviews lasted two to three hours, started with a group interview around the kitchen table and then individual interviews as family members engaged an Internet session, commenting on the people they communicate with and web sites they visited. This paper is also based on interviews with 11 families in the Boston area in 1997, conducted by the first author. They were designed specifically to examine the location and use of the home PC by different members of the family. All families owned a multimedia PC and had children living at home, but represented a spread of income levels (between $20k-100k+ per year), housing types (private house, condominium, apartment) and locations (urban, suburban, rural). Eight of the 11 families had an Internet connection.

Transcripts of both sets of interviews were coded to indicate discussion of topics relevant to the dynamics of computer and Internet use. The resulting topic collections were surprisingly large for both studies, indicating that families had a lot to say about constituent issues such as the location of the computer, and the way it is shared and managed within the family. In the following sections of the chapter we step through the major findings in this collection as they relate to the groups of questions raised in the previous section. Where necessary, we cite relevant quantitative findings to back up the qualitative analysis. We preserve the same ordering of issues and questions as before, addressing the timing, location and shared use of the home computer in turn.

8.6 Results

8.6.1 Temporal Organisation of Family Computing

Routine Timing

Figure 8.1 shows the pattern of daily Internet Mac use by teens and adults within the HomeNet population. The pattern is dramatically different for weekdays versus weekends. On weekdays when home-life routines are dominated by school and work attendance, Mac use and therefore Internet access is more intensive, and concentrated in the evenings. This concentration is especially pronounced for teens, who use it most frequently between 2.00 and 5.00 p.m., immediately upon returning home from school, and then successively less until they go to bed. In contrast, adult weekday use peaks later at 8.00 p.m., but at a much lower overall level. These peaks correspond roughly to "prime time" TV for children and adults, and lend some confirmation to the findings of other studies

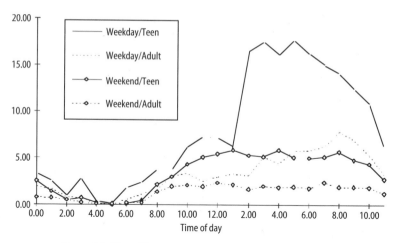

Figure 8.1 HomeNet computer access over a typical day.

that PC time is being taken from TV time. On weekends this prime time effect disappears, with teens and adults using the computer and Internet more evenly throughout the day at much reduced levels.

Within this overall framework, we found ample evidence of regular patterns of individual use. The most routine uses of the Internet centred on the checking of e-mail. As the following quotes shows, this is often done first thing in the morning after waking up or when returning to the home after school or work. Each quote is attributed to one of the Pittsburgh or Boston families by a reference number. Speakers in the Pittsburgh corpus are identified by initials, while speakers in the Boston corpus are identified by their role in the family or interview (M = Mother, F = Father, S = Son, D = Daughter, I = Interviewer).

Pittsburgh 14

BK: I get up, I turn the computer on and then I go, while it's heating up, I go and put water on for tea and then I call up my macmail, which is usually. . .
LW: Six or eight messages, all from her boyfriend . . . laugh . . .

Boston 10

F: I usually around seven in the morning I'll check e-mail between 7.00 and 7.30 and then I will go to work and then when I get home at about 7.30, 8.00 I'll usually go on and design a couple of ads on publisher and then I'll close up around 9.00–9.30 and usually check the website to make sure its up and running because its been crashing a lot and then I shut it down about quarter to ten and that's me. The weekends I try to stay off it just because I don't want to see it.

Boston 6

F: In the evenings I come and check my e-mail and probably sometimes to do a translation um quick translations from a few works or um on the weekend at least four hours on the weekends to edit an article . . .

I: Right does that vary in the day when on a Saturday or Sunday?

F: Sometimes usually

M: Usually do it early

F: Saturday mornings.

As in the statistical data, these routines can be seen to be sensitive to the day of the week. For example, in the last quote above, the father refers to a routine of doing e-mail and short pieces of work on weekday evenings but a longer piece of work at the weekend when there is more time and opportunity. The fact that he chooses to do this task on Saturday mornings rather than at any arbitrary time of the weekend, also reveals an attempt to constrain the amount of time spent on the activity and its impact on family life. Individual routines of this kind are very idiosyncratic and not adequately reflected in the overall trends of Figure 8.1. Thus although this father works on Saturday morning, other fathers avoid PC use at the weekend (as in the second quote above) or use it to play games and relax (quote below). This variation is not captured in Figure 8.1 by the steady but lowered use of the Internet by adults on a weekend morning.

Boston 2

F: On Saturday morning or Sunday morning if I come down and make a pot of coffee and I'm waiting for it to perc I might play a fast game of bridge just cos I'm waiting for the coffee pot to perc through.

Most individual routines for PC and Internet use were designed to fit with those of other members of the family. Thus each family was found to have its own complex set of routines for taking turns on the computer. These were not described in terms of a simple schedule of time slots and users, but rather as a system of turn-taking rules with some typical outcomes. The following quote captures this attitude exactly, and outlines some characteristic patterns of use in many of the families we spoke to:

Boston 5

I: So when would you use it?

M: Its almost always in the evening after dinner especially in the summer. We haven't actually used it as much in the summer 'cos obviously it's nice out and we want to be outdoors. But you know through the year we usually notice it's like I said after dinner. I'll come in, the kids will usually use it first because they're anxious to get on it like right after dinner. They want to come in and get on it and then sometimes they'll get to the point where they're all taking turns on their games and I'm anxious to get done whatever it is I have to get done or whatever, but I wait. So I'm usually later on in the evening. Claudio uses it more during the day because he works off shifts so he has the opportunity when no one's around to hop on and do his cheque book or whatever. So we all use it at different times mostly at night, and I use it mostly once the kids have got settled and they're having their baths and getting ready for bed. I'll come in and work on it at that time.

Many family routines varied not only by day of the week but by seasons. School vacations were particularly significant for both parents and children. The relaxation of school schedules and activities meant that PC and Internet access could be spread more evenly throughout a weekday, and the lack of homework liberated more time for children to play PC games! However, the fact that children spend more time at home during vacations, affected parents working from home:

Boston 2

M: But see we don't separate necessarily how can I say this we work sometimes at our office sometimes here and we are more productive at home and during the school year we actually work more at home

I: Right

M: Because during the summer Becky is here a lot and she does not understand the nature of our work and wants to chat so we have to go to the office a little bit more so we can get things done. But the office is a hard place for us to work – its very busy very noisy.

The extent to which computing routines had become established in family life was revealed by reported reactions to disruptions of various kinds. Going away on vacation or having a computer break down often led to what can only be described as withdrawal symptoms. These symptoms ranged from a heightened sense of appreciation for the PC, to an almost animal-like series of visits to the place where the PC used to be! The addiction to e-mail was so strong in one family that it had led them to seek a public Internet access point on vacation:

Boston 4

M: I really enjoy it. I miss it so much where it's broken down I really enjoy it

Pittsburgh 12

MK: It's pretty useful, since the computer's been in for I guess this little updating and our printer is in here for a repair, I sit in the family room which is adjacent to the living room and I'll be reading the newspaper and watching TV and I'll see the kids keep coming down to the desk where the computer was and then they stop. And they're, it's like if your car is gone and you keep going outside to drive somewhere and you just, they're just stuck. They keep going to this space and there is nothing there for a few days. And I guess if we never got it back they'd quit doing it, but it's kind of funny watching them go for it and it's not there.

Pittsburgh 10

BK: We went down to North Carolina the outer banks for two weeks, my niece and I we just we couldn't stand it we had to go find a computer . . . laugh . . . I mean not to be able to check e-mail you know especially, I mean the chats well I can handle that, but not to be able to check e-mail it was like I couldn't stand it. So we went, we found a library that had, and we stood in line and waited. Of course, it was a small library, they only had one computer you know . . .

Developing Routines

Routines do not emerge full-blown as soon as a household gets a computer, but develop over time, with personal experience and mutual accommodation among household members. Generally, when an individual performs a behaviour repeatedly in similar circumstances, the behaviour becomes internalised and automatic. With practice and repetition, the cognitive and motor activities needed to initiate a behavioural sequence and then execute to completion becomes automatic and performed in parallel with other activities, requiring minimal allocation of focal attention (Schneider and Shiffrin, 1977; Ouellette and Wood, 1998). The behaviour becomes integrated into a larger chunk size. For example, when a person first uses a home computer, each step in booting it up and starting the program for checking electronic mail must be thought about separately. Aiming a cursor with the mouse or typing the return key after entering a form are conscious actions. With experience, however, this action sequence is encapsulated into the higher-level task of "checking my e-mail" and is performed with minimal attention to the details. Not only is habitual behaviour performed in a single, automatic sequence, but the sequence is often set off unthinkingly by environmental events (e.g., the ringing telephone sets off the sequence to answer the phone) or schedule (e.g., finishing dinner may trigger TV viewing). As a result, these routinised or habitual behaviours become highly predictable. In contrast are what might be called "controlled" behaviours, which are directed by intention through deliberate reasoning processes. These controlled behaviours are likely to be performed more slowly and are less stable, with more variability from one opportunity to perform it and another.

In summary, when people first get a new technology at home, they slowly develop routines, which ultimately lead to the highly regular patterns of use we've just described. We examined this process of routinisation by tracking the month-to-month consistency in the times during the day participants in the HomeNet trial used the Internet. We expected to see that this month-to-month consistency in their schedules would increase as they became more experienced in using the Internet.

We first calculated the number of minutes per hour of the day that a participant used the Internet, averaged over a four-week period. Call this vector of 24 averages the participant's Internet schedule for that period. The similarity between an individual's Internet schedules across adjacent time periods is the Pearson correlation of these vectors, with each correlation based on an N of 24 time slots. A high correlation implies that their Internet schedule was similar for two months in a row, while a low correlation implies that one cannot predict when they would use the Internet in one month from their behaviour in the preceding month.

We expect that the average month-to-month correlation would be substantial and that they would increase with a participant's experience online. In this research, we define online experience as the cumulative time that an individual has spent online (i.e., the total number of hours the participant had been online since the beginning of the trial). This metric is correlated with the number of months an individual has subscribed to an Internet service, but weights these months online by the amount the subscriber used the Internet during the month. Thus our measure of experience is behavioural, and does not simply reflect the passage of time.

Figure 8.2 show the average month-to-month consistency correlations in Internet schedules plotted against log to the base two of cumulative hours online. The analysis uses a mixed linear model to predict the consistency correlation based on the participants' gender and adult status, the number of months they have had access to the Internet in their household, and their personal cumulative hours using the Internet. Respondents were treated as a random effect in the model, with an autoregressive error structure of period one. The average month-to-month consistency in Internet schedule was moderate, with a mean Pearson product moment correlation of 0.32. Both the plot and the more formal data analysis show that the month-to-month consistency increased the more participants used the Internet. The coefficient for cumulative hours online means that, on average, as participants increased their time online by a factor of 10, their month-to-month consistency correlation increased by 0.056. An examination of Figure 8.1 shows that this increase in consistency with experience had a steeper slope after participants logged 100 hours online.

Ad Hoc Timing

In addition to using the computer at regular times, people also reported a more spontaneous or ad hoc use. This was often triggered by the need for a particular piece of information or simply finding the PC unattended when they expected it to be in use. Typically, these spontaneous sessions were short and sweet:

Boston 2

M: Um in the evening we use it as people call in and we need to get into the Database to see what a phone number might be.

Boston 10

M10: I'll use it when David will call me and tell me to check on something that's when I usually pull it up or to do something.

Boston 5

I: So you're doing it during the day so you don't have to use it at night when the others want to.

F: Sometimes at night after a meeting or something and I'll want to e-mail something.

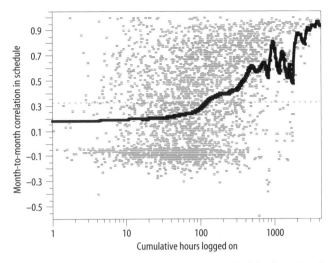

Figure 8.2 Month-to-month consistency correlation in schedule plotted against cumulative hours online (logged). Fixed line is a smoothed, spline fit.

Pittsburgh 10

DH: No, usually if say like my Dad uses it, he'll use it, whatever he does, sign on, work or whatever, then he'll shut it off. Then later I'll see there's nobody on it and I'll turn it on. So it's on and off a lot, usually . . . I'd come on here to just check for e-mail, or like use Netscape and just browse around golf pages or cat pages.

Checking for e-mail was a common ad hoc behaviour. Sometimes this was done during someone else's session by asking them to check the inbox. As in the following quote, seeing or hearing someone logging onto the Internet might be a trigger for this kind of request:

Pittsburgh 4

RK: Show me how you would log on to e-mail.

DB: All right.

(logging on noises)

SB: Whenever anyone does that, he's like "can you check my e-mail?"

DB: Yea, whenever I hear that going I'm like, "Hey can you check my e-mail if you're on there?"

Time-saving Practices

Because time on the PC was generally a scarce resource in the households we visited, individuals had evolved a variety of time-saving practices within and across sessions.

Within sessions, they would sometimes multi-task to make use of one program in the time taken for another to operate. A typical example was listening to an audio CD while backing up data, or checking e-mail while software downloaded. TV watching was also reportedly done in parallel with PC use. Teenagers seemed to have the greatest propensity to do this, even in tasks that apparently don't need much attention like playing games or doing homework!

Pittsburgh 4

DB: It depends what I'm working on. If I'm doing something I really need attention with like if I'm editing resources or something. I want to focus on that or else I can screw up the program. But if I'm working on an English assignment I can type and listen. If I really need to focus on an assignment for school I'll turn it off. But when I play games or something I have the TV on. Or if I'm working over there and I'll have the TV on.

Across a number of sessions, people would organise their tasks according to how much time they had to do them. For example, e-mail processing and web browsing was sometimes done across two sessions, with the first session used to read and filter material and a second longer session used to process and respond to it. Note that printing is referred to in the case of web browsing below, and constitutes another time-saving measure in its own right.

Pittsburgh 10

DH: If I do web crawler or yahoo or something, it'll be like, I'll look for say Monty Python then like if it's something I want to go back to I'll leave you know a bookmark, maybe. If I think of it. I'd go through here, maybe print it out, or download it, or you know it never you know consists of spending very much time with it.

These measures reflect a very sophisticated capacity to estimate how much time is needed for different computing activities and to match this with the amount of time likely to be available on the current session. This kind of calculation was described explicitly by a number of interviewees, and is all the more impressive against a backdrop of multiple users competing for a single shared resource:

Boston 3

F: Sometimes I'll be on for doing something like this (poster) for 10 or 15 minutes you know to revise it but if I'm doing book keeping which is about once a week I'll be a couple of hours.

Boston 7

M: For example, I have to write a memo to another doctor. I'll probably just do it there (at work). I'll find 45 minutes. But if I want to write a more thoughtful kind of memo I wouldn't have the time there. I would have to take it home and do it.

8.6.2 Spatial Organisation of Family Computing

Choice of Home Computer Location

Where computers were located within the home influenced how they were used. Their location in turn is influenced by a number of factors, including the size of the home, the presence of children in the household, whether any household member ran a business from home, and the family's beliefs about the appropriateness of computing technology in various rooms. Figure 8.3 shows the location of the 108 computers in homes.

To understand the choice of locations represented in Figure 8.3, and their effect on home computing we turn now to the interview data. We begin with a review of the problems people associated with different locations, and go on to consider their comments on social interaction around the computer itself.

Location Problems

In general, there was a spread of locations chosen for the computer and an ambivalence about the suitability of all of them. There was little agreement within or between families as to where the *best* location for the computer was. Indeed each location tended to be good for some members of the family but bad for others. This was particularly true of locating the primary computer in a private room of the house such as a child's or adult's bedroom. If it was in an adult's bedroom, the children couldn't get access to it as much as they wanted and if it was in a child's bedroom, the adults couldn't use it when the child had gone to bed. The following quote illustrates this dilemma.

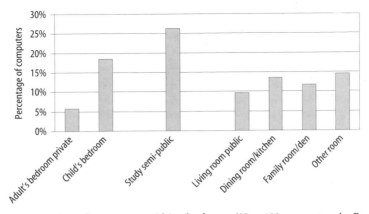

Figure 8.3 Location of computers within the home ($N = 108$ computers in Boston and Pittsburgh).

Boston 4

M: I had it in my bedroom here and after they went to bed I used to go in there and I'd use it. And then I moved it from my room into their room. They [the children] said you had it long enough. You bought it for us.

I: So did that mean that you couldn't use it again?

M: No, I would just go in their room and use it.

I: What even when they were asleep?

M: Oh no, I couldn't use it when they were asleep.

I: So you had to change when you used it?

M: Right, right. I have to use it in their room in the day time when they were in school instead of the night time when it was quiet. So I never get the house-work done during the day.

As a result, only 25 of the 103 (24 per cent) computers in the sample were located in a private space – a parent's or child's bedroom. This placement is surprising, in part, because so many of the families in this sample got their computers for their children. This motivation to get a computer for children is consistent with national data in the USA showing that households with school-aged children are more likely to have a personal computer than households without children (US Department of Commerce, 2000). Families were more likely to place the computer in public spaces like the dining room, kitchen, family room, spare room, or basement (50 per cent of computers) or in a semi-private space, like a study, which had an adult owner, but could be used by all household members (26 per cent of computers).

However, placing the computer in a completely public room such as a kitchen or family room didn't solve these problems either. Although this made the computer equally accessible to all family members, it did so at the expense of privacy and concentration. This made it difficult to use the computer for tasks like e-mail, finances or word processing that require a degree of peace and quietness:

Boston 4

I: OK, so where would you do the games?

M: Probably in the living room, and typing I would do in my bedroom where its quiet and personal and I cannot be disturbed.

Many parents in the sample, however, selected a public place precisely because it denied privacy to their children, as they used the Internet. As we discuss below, by placing the computer in a public place, parents could casually inspect what their children were doing online. As they walked past, they could see what was on the screen, for example, and ask questions about their children's behaviour. Some parents used the public location of the computer as a deterrent, believing that their children would be less likely to visit sexually explicit websites or converse with strangers in chat rooms if their behaviour was subject to parental oversight. Conversely, children lobbied to have the computer placed in their rooms because of the privacy it afforded them.

Pittsburgh 20

EP: Carnegie Museum is a wonderful place, but I wouldn't leave him alone with a map in the middle of it. So it's just kind of parental supervision . . . I mean we're in the same room but its just sort of knowing when he's on . . . I'd be sitting on the sofa knitting or watching.

One compromise was to locate the computer in a semi-private but shared room, such as an office. This made it more accessible to all the family but capable of private use when necessary. However, even here, there were problems with ownership of the computer falling to the father of the family, and the feel of the computer being too work-oriented. In larger homes, there were also logistical problems with moving the computer too far away from the hub of family life. If it takes too long to walk to the computer, switch on and connect to the Internet, then a more spontaneous and sporadic use of e-mail or the web is rejected by families:

Boston 11

F: You'll see when you go downstairs (office) you're in a different mood you're not relaxed like you are up here (family room).

Boston 11

M: I get tired of going downstairs and all of a sudden I think gee I'd better e-mail Lauren in Singapore, so I have to go all the way downstairs, and basically I live on this floor because I'm doing the dishes . . . Its just like people build and they put the washer dryer on the second floor so they don't have to go all the way down to the basement to put the clothes in one machine.

Pittsburgh 9

MTR: I would e-mail people and say . . . just pick me up at the airport, you and me, call me on the phone and tell me. Because if you send it e-mail, who knows when I'll be up here to read it again? So, I would e-mail people and tell them to telephone me. Because I wasn't going to hiking up to the third floor to get connected, you know, on the chance that something could be there or not, so that's it. If it was something I needed to know I would send the e-mail and say call me.

All these problems show that the simple choice of where to locate a computer in the home has large effects on family life, both in terms of the way individuals use the computer and also in terms of the way they share their time on it. These problems appear to change rather than diminish as multiple computers enter the home. While sharing becomes less of a problem, control and interaction within the family becomes more difficult. This is illustrated in the next section, which deals directly with the effect of home computing on social interaction within the household. As we shall see, this is not all bad news as both sociofugal (separating) and sociopetal (combining) effects are apparent!

Sociopetal and Sociofugal Effects of Home Computing

Just like the placement of chairs around a table, the placement of PCs around a house appears to have consequences for social interaction among its users. In general, the PC seems to be a sociable device, somewhat akin to a table or a television in bringing people together around a common activity. This *sociopetal effect* was indicated by the very large number of reports of joint PC use in both sets of interviews. In some cases, the encounter was described as being similar to television for at least one of the parties, who might watch another person's interaction with the PC while waiting for their own turn on the machine. This of course provides an opportunity for vicarious learning of interfaces and applications, which can be applied later on. However, even in these cases, the watching may lead into a more active involvement with the interaction, through discussion and direction that goes beyond the television experience:

Boston 5

M: Sometimes they're watching me. Sometimes Ewan and Roger will come in if I'm working on a project whether its on the Internet looking at something in particular they'll watch me, or if they're interested in what I'm doing with work or whatever, or sometimes they'll just be waiting for me to get off. Or they'll sit there, they'll discover something and they'll be like "Mom mom" you know, and I'll come in and I'll sit down and Ewan will sit down and we'll watch Roger or something with this great discovery that he's made, whether its a city he's building or something he's found on the Internet. So we'll just watch. It's a way to interact and do something together which really goes beyond what you can do with the television.

The ability to watch or be called over to view someone else's PC session is clearly increased when the PC is sited in the public rooms of the house. However, it also depends on the type of activity being performed on the PC by the primary user, and can happen in the most private of spaces. For example, the quote above applies to the use of a single family computer located in a corner of the parents' bedroom. Sharing a computer is difficult at a close viewing distance with single-user input controls. Compare this to the experience of using a games console with multi-user controls and a TV screen about nine feet away.

A wide variety of local applications were cited in the reports of shared PC use. Games were the most commonly mentioned, and included parents playing with children as well as children playing with siblings or friends. Other applications that seemed to bring people together were creative activities like making movies or cards. Even very personal applications could bring people together when one person was teaching or helping another:

Pittsburgh 4

DB: We'd make up jokes like that. And wasn't really cause we wanted to make it a comedy, because well its just fun on the nights we have sleepovers and record stuff.

SB: It keeps them off the street corner basically (laughter) . . . I come down 3 o'clock in the morning and a kid, cornstarch in his hair, dancing around in front of here. And my kid is up there with a camera. That's a lot of fun.

Boston 10

F: I was the one that taught Carla how to do the invitations.

Boston 2

M: Carrol and Becky learned how to type by using Mavis Beacon – they learned together.

Internet applications were even more effective than local applications in fostering social interaction around the computer. This can be seen statistically from the reports of joint computer use after 9 months in the HomeNet trial. One-third of all sessions were reported to be with others, and 75 per cent of these sessions involved Internet use (see Figure 8.4).

Searching the web together was often mentioned as a joint PC activity. Sometimes this was done as a conscious joint activity from the outset, while at other times people got drawn into doing it together as a result of being called over to see a piece of interesting content. For example, one married couple in the real estate business used to enjoy regularly "cruising the world", looking at expensive houses together. In another example, a daughter showed her mother how to print out route directions for guests attending a family reunion. Such sessions were generally

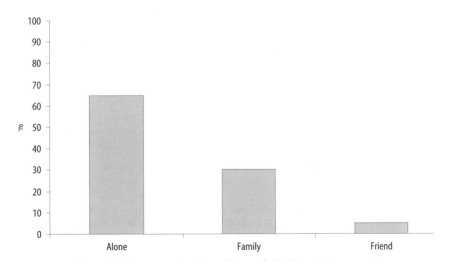

Figure 8.4 Reports of co-located use of the HomeNet computer.

seen in a positive light, as occasions that enhanced family relationships and time. This can be seen most clearly from the following quote describing the discovery of Santa Claus's homepage:

Pittsburgh 6

SK: How has that affected your relationships with each other?

RC: Well, it was interesting we just happened to find Santa Claus's web page. And it sounds ridiculous, but we spent an hour together as a family. We typed in each other's, each one's name, and they give you whether you've been good or bad, and then they say, yeah well, what you've done. And the nine year old she didn't believe in this and then it said, you should be neater, and she went, how did he know! . . . laugh . . . It was just a lot of fun. And then they had a quiz, and you got your elf diploma, you print it out and it's signed by Santa Claus. So it was really a good thing for the family, for young kids. We just had a good time with it.

There were fewer reports of joint e-mail or chat sessions on the Internet. Communication appears to be a more personal and private computing activity than information access. Perhaps for this reason, when shared communication behaviour was mentioned it was characterised as a particularly intimate thing to do. This is indicated in the following quote from a daughter who regularly helped her mother compose chat group messages:

Pittsburgh 14

BK: And on the chat groups a lot of them know my mother and she sits there and talks through me. You know I type what's she saying 'cos she can't type. So it's actually brought us closer. You know we have more conversations now, because it's going through to somebody else.

Despite the beneficial effects of the PC in bringing family members together, there were serious concerns about more long-term *sociofugal effects* of keeping individual members apart from the family. These concerns were usually expressed by parents in the context of talking about the growing isolation of their children. The following quote is typical of these concerns since it mentions the relatively large amounts of time children and teenagers can spend on computer games when the parents are out of the house or busy with other things. In this example, the presence of the computer appears to affect the family time spent by a son with his parents, and also the playtime spent with a visiting friend:

Boston 7

M: It's funny because sometimes I feel like it becomes a solitary thing for Steven up here. He could spend 2 to 3 hours and to me that's like, doing this for 2 to 3 hours is too much and I don't like it. And then his friend Andrew came round today. And I told his mother "Tell Andrew there's no computer in the house today. Someone was bad and it's gone". Because he's the kind of kid that will come over and solitarily do something. And then they won't play. That's OK with

> Steven because he can do something solitary too. But the point of playing is to play together, to do something . . . To me its like, "What kind of impact does this have on your kids?"

Parents also recognise the potentially antisocial nature of their own computing behaviour, particularly when they share their children's passion for games. Again the overall amount of time spent on the computer, in relation to other activities is seen to be a key factor. However, the fact that they can articulate and discuss this concern, shows a level of insight into this effect that the children do not have:

> *Boston 6*
>
> F: What do I think of computers?
>
> I: Yeah
>
> F: They are very useful. They are, um you know, there is this almost like they have this city inside of them.
>
> M: A world.
>
> F: And um I can get my work done and be entertained.
>
> M: They offer a lot but as long as you know when to put the brakes on. Because you could spend your whole life, day after day I mean, I could I always say its a good thing. I don't gamble because I have such a hard time tearing myself away from something like this . . . I get on a game late at night and I probably won't go to bed till 2 in the morning. I mean the idea is that you can get your work done faster and then go enjoy life, but really what happens is you can do so much more that you do so much more – d'you know what I mean?

As a result of these and other concerns, parents try to constrain their own home computing behaviour and that of their children. Exactly how they do this is explained in the next section, together with the attempt by children themselves to reassert their rights to the computer through increased expertise.

8.7 Power, Regulation and Control

8.7.1 Parental Regulation of Computer Turn-taking and Internet Access

In a prior section we saw that families develop routine patterns of turn-taking at the computer, as a way of dealing with contention for computer time. What was not so clear from that section was how such patterns come about, and what happens in cases where the routine practices break down with individual violations or shifting demands. We briefly consider these issues here, since they relate to a significant power struggle for computing resources in the home. This is effectively part of a bigger power struggle between parents and children to structure and manage family life itself. It is important to understand this battle, since it lies at the heart of the social context for home computing, and cannot be overcome by simply increasing computing resources and locations in the home.

Contention for computer time is a heated issue in many of the families we visited. Families do not sit down calmly at the beginning of the week and schedule time slots together. According to our informants, they watch the space in which the computer sits, try to read each other's plans, and fight for a seat:

> *Boston 4*
> M: We'd get into a fight.

> *Pittsburgh 12*
> MK: They fight over it like they used to fight when we only had one TV.

> *Boston 9*
> M: I've seen people literally pushed off that chair.

> *Boston 5*
> M: I wouldn't say we have a problem with conflict but it does arise just in the manner of seven of us using the same computer.

Given this situation, it falls to the parents to arbitrate and ensure that everyone in the family gets a "fair" amount of time on the machine. Parents do this in different ways. Some parents allocate time limits to stop the dominant children from taking too long. Others enforce sanctions if the children can't agree to sort it out themselves, or negotiate on the basis of who needs it most. In general, school or homework takes priority over recreational uses, and whoever goes to bed first tends to get the earlier time slot:

> *Boston 5*
> F: When they're playing the games we set time limits so everyone has a turn.

> *Boston 4*
> M: What I do is I say "OK nobody will use the computer. We will decide who needs it and which is more important".
> I: Yeah, so it goes on who needs it the most?
> M: Right, who needs it the most. If it's to play a game then no. Then if it's to do school work then fine then he gets the priority.

> *Boston 11*
> F: My son gets priority because he goes to bed earlier. She stays up later so she can have it later.

In addition to arbitrating between family members for time on the computer, parents also regulate children's overall access to the Internet. Most parents could relate stories of inappropriate content coming up in response to web searches and were wary of leaving children unsupervised on the Internet. Others expressed a general distrust of chat rooms or e-mail. A common metaphor was to liken unsupervised Internet use to leaving young children alone in a public place:

Pittsburgh 20

SP: What is it about him having access to it himself that makes you nervous?
EP: Well, in terms of the World Wide Web I guess you know I wouldn't leave him downtown by himself and say you know here's the number of your bus find your way home. I mean he's smart for his age, he started reading when he was three. But still, he's not so savvy that I'm comfortable turning him loose that way. But with the web it's more . . . You know the Carnegie Museum is a wonderful place but I wouldn't leave him alone with a map in the middle of it either.

These reservations often led parents to ban Internet use to pre-teen children altogether, or to limit and supervise their access. These attitudes softened for teen use of the Internet, but did not disappear entirely. While teenage children were generally allowed access to the Internet, this was usually according to a strict set of instructions by parents and was subject to monitoring and punishment. In some cases, parents had resorted to a form of spying on their children by reading over their shoulders or logging on under their user name to read personal e-mail messages:

Pittsburgh 14

BJ: Freida, do you know what she means when she talks about muds?
FW: Oh yeah, I've sat and read behind her you know what's been going on and stuff like that. I try to monitor a little bit, because she is you know a minor, and all the things they talk about on the computer. And I'll read over her shoulder and go, what's that mean, what's this?

Boston 3

M: Every once in a while I'll read one of her e-mails from her rent people and see what's going on.
F: Yeah, I'll do that to but I don't tell her that though.
M: I think she knows. I don't think she really cares. I don't know if I'd want it to be totally private.
F: She isn't crazy! She deleted all of the outgoing messages every one of them because she didn't want us to read her outgoing messages.

An additional consideration for some families was the cost of a dial-up Internet connection. Parents would oscillate between trusting their children not to connect for too long, and banning use when that trust is broken. Not surprisingly, this leads to an atmosphere of deception and mistrust:

Pittsburgh 4

DB: . . . It was funny.
RB: Until you had a $115 AOL bill maybe. And we just said that's enough of this.
(general laughter)
RB: . . . That was it. That got shut off real fast.

Boston 11

F: But when he's typing and we come down and we find out he's playing on AOL so we have a yell and a scream session and that's the end of that.

In short, a variety of rules and regulations are developed and administered by parents to control their children's access to the computer and the Internet. These rules are designed to ensure a fair distribution of computing resources within the family, based on the age and need of family members. Routine practices emerge from this process insofar as the rules and conditions allow. However, these are always subject to revision and re-negotiation, and can be swept away in the face of an urgent need for the computer or an external family event.

8.7.2 Child Control of Computer Settings and Expertise

Despite attempts by parents to constrain their children's computer and Internet use, children have more free time than adults and a more playful and experimental attitude to the technology. This means that children may actually end up spending more time on the computer than their parents, and will try out things for fun rather than to get some task done. For example, many children told us about changes they had made to screen settings, icons and file systems in order to personalise the computer. They also reported downloading software from the Internet, adding bookmarks and addresses and generally performing a variety of system administration tasks. Because most systems we encountered were not carefully partitioned and managed via multiple user names, these changes affected everyone else using the computer and were perceived to be disconcerting or annoying by other siblings and parents:

Pittsburgh 14

BK: I have that with my niece, she likes to download pictures. I never know what's going to be on the screen.

Pittsburgh 8

MAR: It seems that every time I have mine on here, I don't know what happens to them. I don't know if you can erase them and that's what my brother does to me, but like I had all my college ones on here, and I think he just erased most of them.

Pittsburgh 19

GH: I think she captured Netscape 3 and we had problems with that. And I questioned whether or not she was taking it off the Internet, whether it would have bugs or anything but she ran a de-bugger program and found one mistake and reloaded.

One effect of this kind of playfulness is that children and teenagers become more competent and knowledgeable about managing the computer than their parents. Teenagers in particular were very adept at using the computer and solving technical problems. This meant that they often became the technical support gurus of the family, and would be consulted by their parents and younger siblings about technical problems and goals (see Kiesler et al., 2000 for further details of this phenomenon). Both generations acknowledge this role as the following quotes show. Note also that the son referred to in the third quote below has left home, but still acts as a system consultant to the family!

> *Pittsburgh 19*
> JH: My brother is like the director of the house.
> BJ: I see.
> JH: I'm second in command.

> *Pittsburgh 19*
> JH: He taught me a little bit and I just found out the rest on my own. I'm basically a trial and error person. I learn a lot of things by myself, I don't like to sit down and listen to people telling me how to do stuff unless I know I have a problem in a certain area, and my Dad just doesn't know. It's tough to explain it to him because he's not used to it at all. Totally different generation.

> *Pittsburgh 16*
> RC: It's embarrassing because my nine-year-old granddaughter does better than I do.

> *Pittsburgh 16*
> JH: When he comes home . . . then we usually have a couple of questions for him as to you know, why is this happening and you know. He seems to have all the logical information as to what's going on. He's our source. The house source.

This asymmetry in knowledge about the computer is significant in the context of the power struggle between parents and children for computing time and access. It leads to an unusual social situation in which the normal power relations are partially reversed. Parents have the power to veto or limit access to the machine, but children have the power to modify its set-up and operation.

8.8 Discussion

These findings go some way towards unpacking the social context of home computing, at least for a small sample of American families struggling to accommodate yet another piece of technology into their lives at the end of the millennium. Whereas local PC applications formed the basis of computing activities at the beginning of the 1990s (Venkatesh, 1996), Internet services have now added to the functionality and appeal of the PC, providing something for everyone in the households we visited. However, services had *not* taken over from local applications, but rather increased the mix of local and remote software and content used on the same device. Viewed from the user's point of view, the difference between "local" and "remote" was irrelevant to the tasks they were carrying out, except where it affected task performance. For example, a decision about whether to use a CD-ROM encyclopaedia or an educational website for a piece of homework would probably hinge on factors like the speed of access and the quality of information, rather than on some overall preference for or against the Internet. Furthermore, because the point of access is the same for local and remote information, the social issues of turn-taking and timing, spatial location and control apply equally to both dimensions of computing. This means that in households where the primary Internet access device is a computer, a person's overall Internet experience is part and parcel of their home computer experience, and does not depend on Internet service offerings alone. Indeed as we have seen, it depends as much on how many people have to share the computer, what place they occupy in the household, where the computer is located in the house and whether they are allowed to access Internet services at all!

A convenient way of summarising these contextual effects is shown in Table 8.1. This contrasts our findings on the local adoption of the home computer with known findings on the adoption of TV (e.g. Gunter and Svennevig, 1987). We have chosen the TV as a point of reference because there are many similarities in the use of the TV and PC, but also significant differences which highlight the PC's distinctive role in family life compared to its more familiar cousin. In order to return to the research questions that motivated our study, we have divided the table and findings by the major contextual factors they relate to. Hence, we step through findings on the temporal and spatial organisation of computer use, and on its relationship to social interaction and control. After reviewing these findings shown in the table, we go on to consider their implications for the design and marketing of computers and other Internet devices in the home.

Regarding the timing of PC use we found that it clustered within the same time periods as "prime time" TV use. Hence weekday evenings were the most popular time of the week for using the computer and television, as family members return to the home after school or work and settle down after eating. These findings also indicate where the majority of PC time is coming from, within the existing commitments and behaviours of individuals. It is often taken directly from TV time, as indicated in the large-scale market research and time use studies. However, whereas the scheduling of TV use is driven largely by the programmes on offer, the scheduling of PC use is based on personal schedules and patterns of turn-taking within the family. Both kinds of schedule lead to repetitive and routine behaviours, but these are subject to greater negotiation and revision on the PC where the content is open-ended. This also reflects the fact that PC use is primarily a personal activity, even though it might

Table 8.1. Contextual factors in the adoption of the home computer compared with television

Context	Television	Computer
Timing of use	Prime time	Prime time
	Routines stemming from programming schedules	Routines stemming from personal schedules and time sharing patterns
	Reactive use	Reactive and opportunistic use
Spatial location	Solitary/1st TV- Public family room	Solitary/1st PC – Semi-public office or private adult bedroom?
	2nd TV – Private bedroom	2nd PC – Private child's bedroom or spare room/Public kitchen or dinning room or family room?
Social interaction	Conversation	Conversation
	Shared presence	Support
		Collaboration
Control	Parental arbitration of time and content	Parental arbitration of time and content
	Based on interest	Based on interest, need and cost
		"Child" maintenance and repair

come to be shared by others along the way. There is therefore a greater sense of ownership of individual "sessions" on the PC than on the TV, with one person allocated overall control. In addition to scheduled time at the TV and PC, family members also engage with them more spontaneously. People may switch on the TV to "see what's on", or notice a programme that someone else is watching. In the same way, they may see something of interest on the PC over the shoulder of the current user, or get called over to help, or find that they have e-mail waiting to be read. This kind of *reactive use* of each device is supplemented on the PC by a sheer *opportunistic use* resulting from finding it free. Children in particular may slip onto the PC in this way, to overcome time sharing constraints before a fixed bedtime.

Table 8.1 also shows the differences between the spatial location of TV and PC use. In both cases a key factor is the number of devices in the home. When there is only one TV or PC in the household its location is chosen differently from when there is more than one. The location for a solitary TV is often the family room while the location for a solitary PC is often the home office. Subsequent televisions may be placed in more private rooms of the house such as a bedroom. However, the placement of second computers is less predictable from our data, which confounds form factor, age and Internet capabilities, at least within the HomeNet families. All we can say is that second computers turn up in a variety of rooms within the home, which may be private as in a child's bedroom, or public as in a kitchen/diner or family room. A significant factor in the choice of second PC room location is likely to be its status along a *work-play* dimension. Both functions are evident in the use of a solitary PC, but appear to separate somewhat with the introduction of a second PC in the home. Typically the first PC may remain in the office as a work-oriented machine while the second PC becomes more specialised for recreation. In this scenario, the second PC might be located in a more recreational room to match its function. This contrasts with the situation today with the TV, which is almost exclusively used for "play", wherever it is located and however many sets there are in the house.

The question of whether or not the presence of a computer in the house brings families together or pushes them apart, is addressed in the third row of Table 8.1. As with the TV, the home PC gives people a common basis for conversation within the family as things come up which match common interests. However, whereas the intensity of interaction around the TV is low, and characterised largely by co-presence in front of the set, the intensity of interaction around the PC appears to be higher. Family members may enter into true collaborations with each other to operate a PC programme or Internet service together. Also, the fact that the PC is difficult and unreliable to use means that family members offer or solicit support from each other in a way not found with the TV. These kind of sociopetal effects of the TV and PC are probably greatest in public rooms of the house where family members are already in close proximity to each other, and with solitary devices whose use is not diluted by the availability of other models.

Finally, we have found that PC and Internet use at home is controlled largely by parents. Control applies to the overall time spent on the computer as well as the kind of content viewed within that time. This appears to be similar to the control exercised by parents over TV use. One difference is that PC use appears to be regulated on the basis of interest, need and cost rather than on interest alone. In addition, the growing expertise of children in operating the computer often puts them in a better position than their parents to control maintenance and repair tasks. Again, this adds an extra level of complexity to the negotiations for PC time and access compared to that for the TV. Thus on every dimension, the PC turns out to be an altogether more complex technology and context for interaction than the TV.

These findings on the social context of home computing have a number of implications for the marketing and design of domestic technology. In the case of the home computer, they suggest that it might be better adapted to a *multi-user* context than it currently is. For example, its propensity to stimulate joint activity and collaboration might be accentuated by providing multi-user controls at a further distance from the screen. Certain creative applications might be targeted for this support, together with general web browsing, both of which were found to foster collaboration between family members. A "distant" screen mode might also be used to display a range of content in the absence of particular users. This might be done as an extension of screen savers which can already recycle photographs and other items of interest as a background information channel. Another implementation might be to notify users of the arrival of e-mail or other communications, on the screen or monitor casing. Both facilities would cater for the multiple interests of individual family members, and allow them to time their interactions with the machine a little more intelligently. A further enhancement to the multi-user features of the home computer would be to add timers and history logs, to the existing facilities for user settings and Internet content controls. These could be used quite simply to set time durations for PC or Internet sessions, and allow parents to review session activities at mutually acceptable levels of granularity.

The relationship of one computer to another in the home might also be exploited in the way computers are designed and marketed. The fact that primary and secondary computers come to be used in different ways and in different rooms could be influenced by design. One possible split would be to design "work" and "play" machines for primary and secondary use in the home. Alternatively, computers might be designed for a combination of work/play uses appropriate to particular rooms – such as a child's bedroom or the kitchen/diner area. Another possibility would be to sell portable machines that can be carried between different rooms for different purposes. In every case, the effect would be to acknowledge and support the complex partitioning of devices, uses, rooms and users that currently goes on in multi-PC homes, rather than ignoring it through the release of standard, standalone computers.

The possibility of building computer and Internet functions into existing home devices like TVs or telephones is also raised by this latter approach. Perhaps families would be better off with a Digital/Interactive TV or an enhanced games machine as their second "play" PC. Adoption of the PC is already very TV-like as shown in Table 8.1, and it would be a short step for many families to imagine combining their functionality. Plus, the TV is already designed for the kind of joint viewing and inter-action we have just recommended above for multiple users. Unfortunately we cannot really say from our data whether interactive television will be a success in the long term, despite slow sales in the short term. Table 8.1 also suggests that personal schedules may clash with programme schedules on a TV and overload an already well-used entertainment resource with information and communication functions. This is a good place to finish our discussion since it reveals again the complexity of the domestic context for technology design and use. More research is need to understand this relationship better, and to improve the home computing and Internet experience through context-sensitive design.

Acknowledgements

This research was funded in part by the National Science Foundation (Grants IRI-9408271 and 9900449) and by Hewlett-Packard Laboratories. In addition, HomeNet data collection was supported through grants from Apple Computer Inc, AT&T Research, Bell Atlantic, Bellcore, CNET, Intel Corporation, Interval Research Corporation, Hewlett Packard Corporation, Lotus Development Corporation, the Markle Foundation, The NPD Group, Nippon Telegraph and Telephone Corporation (NTT), Panasonic Technologies, the US Postal Service, and U S West Advanced Technologies. This data analysis was carried out while Robert Kraut was Visiting Professor at Hewlett Packard Laboratories Bristol.

We particularly thank our colleagues at Carnegie Mellon University (CMU) and Hewlett Packard (HP) who helped with the original set-up and fieldwork for this paper. At CMU these people include Jane Manning, Sara Kiesler, Tridas Mukophadhyay and William Scherlis. At HP these people include Amy Silverman, Susan Dray (on contract from Dray & Associates Inc.), Cath Sheldon, Dave Reynolds, and Phil Stenton.

References

Aries, P (1962) *Centuries of Childhood*, New York: Knopf.

Aronsen, SH (1977) "Bell's Electrical Toy: What's the Use. The Sociology of Early Telephone Use", Ch. 1 in I De Sola Pool (ed.), *The Social Impact of the Telephone,* Cambridge, MA: MIT Press.

Bass, W, Green, E and Esselink, AK (1996) *PC Time and Money,* Cambridge, MA: Forrester Research Inc.

Baecker, RM (1993) (ed.) *Readings in Groupware and Computer-supported Cooperative Work,* San Mateo, CA: Morgan Kaufmann.

Betchel, RB, Achelpohl, C and Akers, R (1972) "Correlates between Observed Behaviours and Questionnaire Responses on Television Viewing", in EA Rubenstein, GA Comstock and JP Murray (eds.), *Television and Social Behaviour. Vol. 4. Television in Day-to-day Life: Patterns of Use,* Washington DC: US Government Printing Office.

Birnbaum, J (1997) "Toward Pervasive Information Systems", *Personal Technologies,* Vol. 1, pp. 11–12.

Bower, RT (1973) *Television and the Public,* New York: Holt, Rinehart & Winston.

Brand, S (1995) *How Buildings Learn: What Happens after They Are Built,* New York: Penguin Books.

Csikszentmihalyi, M and Rochberg-Halton, E (1981) *The Meaning of Things: Domestic Symbols and the Self,* Cambridge: Cambridge University Press.

Ehrenberg , AS (1986) *Advertisers or Viewers Paying?,* ADMAP Monograph. London: ADMAP.

Fairclough, G (1992) "Meaningful Constructions – Spatial and Functional Analysis of Medieval Buildings", *Antiquity,* Vol. 66, pp. 348–66.

Friedman, A (1998) "The Next Home: Expanding Housing Choice and Flexibility", *Sociological Abstracts Supplement 182,* July 1998, 98S34340, p. 120.

Frohlich, DM, Dray, S and Silverman, A (2001) "Breaking Up Is Hard to Do: Family Perspectives on the Future of the Home PC", *International Journal of Human-Computer Studies,* Vol. 54, pp. 701–24.

Goffman, E (1959) *The Presentation of Self in Everyday Life,* Garden City, NY: Doubleday.

Gunter, B and Svennevig, M (1987) *Behind and in Front of the Screen: Television's Involvement with Family Life,* London: John Libby.

Hall, E (1966) *The Hidden Dimension,* New York: Premier Books.

Harvey, AS, Szalazi, A, Elliott, DH, Stone, PJ and Clark, SM (1984) *Time Budget Research: An ISSC Workbook in Comparative Analysis,* Frankfurt/New York: Campus Verlag.

Heath, C (1986) *Body Movement and Speech in Medical Interaction,* Cambridge: Cambridge University Press.

Heath, C and Luff, P (1992) "Media Space and Communicative Asymmetries: Preliminary Observations of Video Mediated Interaction", *Human Computer Interaction,* Vol. 7, pp. 315–46.

Hediger, H (1955) *Studies of the Psychology and Behaviour of Captive Animals in Zoos and Circuses,* London: Butterworth.

IBA (Independent Broadcasting Authority) (1987) *Attitudes to Broadcasting in 1986,* London: IBA.

Kiesler, S, Zdaniuk, B, Lundmark, V and Kraut, R (2000) "Troubles with the Internet: The Dynamics of Help at Home", *Human-Computer Interaction* Special Issue, Vol. 15, No. 4, pp. 323–51.

Kraut, R, Scherlis, W, Mukhopadhyay, T, Manning, J and Keisler, S (1996) "HomeNet: A Field Trial of Residential Internet Services", *Proceedings of CHI '96,* pp. 284–91, New York: ACM SIG-CHI.

Lacohee, H and Anderson, B (2001) "Interacting with the Telephone", *International Journal of Human-Computer Studies,* Vol. 54, No. 5, pp. 665–99.

Lawrence, RJ (1987) *Housing, Dwellings and Homes: Design Theory Research and Practice,* Chichester: Wiley.

Levi-Strauss, C (1963) *Structural Anthropology* (trans. C Jacobson and BG Schoepf), New York: Basic Books.

Lipman, A (1967) "Chairs as Territory", *New Society,* Vol. 9, No. 283, pp. 564–66.

Luff, P, Hindmarsh, J and Heath, C (2000) *Workplace Studies: Recovering Work Practice and Informing System Design,* Cambridge: Cambridge University Press.

Lull, J (1980) "The Social Uses of Television", *Human Communication Research,* Vol. 6, pp. 97–209.

Lull, J (1982) "How Families Select Television Programmes: A Mass Observational Study", *Journal of Broadcasting*, Vol. 26, pp. 801–11.

Mateas, M, Salvador, T, Scholtz, J and Sorensen, D (1996) "Engineering Ethnography in the Home", *Companion Proceedings of CHI '96*, pp. 283–84. New York: ACM Press.

Nie, NH and Ebring, L (2000) *Internet and Society: A Preliminary Report*. Stanford Institute for the Quantitative Study of Society.

O'Brien, J and Rodden, T (1997) "Interactive Systems in Domestic Environments", in I McCLelland, G Olson, G van der Veer, A Henderson and S Coles (eds.), *Proceedings of the Conference on Designing Interactive Systems: Processes, Practices, Methods and Techniques*, pp. 247–59, New York: ACM Press.

Osmond, H (1957) "Function as the Basis of Psychiatric Ward Design", *Mental Hospitals (Architectural Supplement)*, Vol. 8, pp. 23–29.

Ouellette, J and Wood, W (1998) "Habit and Intention in Everyday Life: The Multiple Processes by which Past Behavior Predicts Future Behavior", *Psychological Bulletin*, Vol. 124, pp. 54–74.

Robinson, JP (1988) "Time Diary Evidence about the Social Psychology of Everyday Life", in JE Mcgrath (ed.), *The Social Psychology of Time*, Sage.

Robinson, JP and Godbey, G (1997) *Time for Life: The Surprising Ways Americans Use their Time*, Pennsylvania: Pennsylvania State University Press.

Schneider, W and Shiffrin, RM (1977) "Controlled and Automatic Human Information Processing: I Detection, Search, and Attention", *Psychological Review*, Vol. 84, No. 1, pp. 1–66.

Silverstone, R (1991) "Beneath the Bottom Line: Households and Information and Communication Technologies in an Age of the Consumer", *PICT Policy Research Paper* No. 17, Brunel University.

Sommer, R (1959) "Studies in Personal Space", *Sociometry*, Vol. 22, pp. 247–60.

US Department of Commerce (2000) *A Nation Online: How Americans Are Expanding Their Use of the Internet*, Washington, DC: US Government Printing Office.

Venkatesh, A (1996) "Computers and other Interactive Technologies for the Home", *Communications of the ACM*, Vol. 39, No. 12, pp. 47–54.

Welman, B and Tindall, D (1993) "Reach Out and Touch Some Bodies: How Social Networks Connect Telephone Networks", in W Richards Jr and G Barnett (eds.), *Progress in Communication Sciences*, Vol. 12, pp. 63–93. Norwood, NJ: Ablex.

Wheatley, P (1971) *The Pivot of the Four Quarters: A Preliminary Enquiry into the Origins and Character of the Ancient Chinese City*. Edinburgh: Edinburgh University Press.

Design with Care: Technology, Disability and the Home

9

Keith Cheverst, Karen Clarke, Guy Dewsbury, Terry Hemmings, John Hughes and Mark Rouncefield

9.1 Introduction: Sociology, the Home, Design and Disability

> It is known that many products are not accessible to large sections of the population. Designers instinctively design for able-bodied users and are either unaware of the needs of users with different capabilities, or do not know how to accommodate their needs into the design cycle (Clarkson and Keates, 2001).

It is now widely realised that the home is likely to prove an important site for new information technologies (Venkatesh, 1985, 1995; Crabtree et al., 2001). This book documents the extent to which the convergence of a number of technologies that link computers with various communication and entertainment technologies have created new possibilities of home shopping, video on demand, home banking, and so on. At the same time other reports such as the EU report on the electronic home (Moran, 1993) have identified a number of social trends – near zero population growth, the rise of the proportion of the elderly, the decline of multi-generational households, the increased number of "non-traditional" homes, new forms of work, increasing leisure time, etc. – that may prove fundamental in shaping ideas about the development of the electronic home. Until comparatively recently, however, little attempt has been made to systematically study home life, as the EC commented:

> No model of the home or its users has been developed which could underlie developments in the Electronic Home area. The initiatives are largely the result of a "technology push" type approach. A clear conceptual paradigm has not emerged (Moran, 1993).

Although the household is one of the most familiar of social institutions, relatively few studies of the household exist and those that do generally comment on topics other than its everyday organisation and life. There are, of course, historical studies of the ways in which family patterns have changed over the centuries, as well as the extremely extensive anthropological studies of kinship patterns in societies all over the world. One of the main problems in thinking clearly about the household is the number of myths that have to be addressed. Myths about the declining nuclear family, the loss of community, the growth of urban malaise, and so on; myths that are themselves occasionally the outcomes of sociological research, myths that may be attenuated and long-lasting in the case of the disabled. Until the 1980s, and the advent of feminist sociology, there were few studies that attempted to investigate household interactions within the domestic environment itself. In terms of household technologies, in the 1980s a number of studies were carried out focusing on the use of the media, particularly television, and communication technologies and the ways in which they were actively incorporated into everyday lives and conversation (Morley, 1986, Silverstone, 1994). Despite this interest, these studies were generally oriented at addressing large-scale theoretical issues within sociology, such as modernity and alienation, rather than examining households as socially organised phenomena achieved in and through the everyday interactions of their members. What was lost were the details of the household as a socially organised area into which the technologies were placed and through which they found their role within the domestic environment. Recently Venkatesh (1985, 1995, 1996) has drawn attention to the importance of the interaction between the technological and social arrangements of the home:

> From the technology side, this conceptualisation shows how computers and new media technologies may be adopted and used; from the user side, it helps identify the internal dynamics of family life that determine successful (or unsuccessful) adoption and use of the technologies. This dynamic can be summed up as the interaction between the social space and the technological space . . . *We cannot assume that what the technology can do in the household is the same as what the household wants to do with the technology* (Venkatesh, 1985 – our emphasis).

Other studies have taken themes identified in Venkatesh's pioneering work, focusing on aspects of the interaction of technology and home life (Mateas et al., 1996; English-Lueck and Darrah, 1997). Research on the Silicon Valley Cultures Project (English-Lueck and Darrah, 1997), for example, traced the effects of technology in the "mundane activities of everyday life" arguing that "we need to know how the many devices entering people's lives are actually used by real people". In previous research (Hughes et al., 1998, 2000; O'Brien and Rodden, 1997; O'Brien et al., 1999) we have begun to indicate some of the issues of key importance in the design of domestic systems, drawing attention to the social organisation of household routines (Hughes et al., 2000). This consideration has been based largely on the notion of "scoping" design activities in a fairly broad manner – producing sensitising concerns for designers.

Clearly design for this particular setting and user group needs to contend with different myths and fears about disability, the home and family, and about technology. This chapter presents some of the very early design work of the "Care in the Digital Community" research project begun under the EPSRC IRC Network project EQUATOR. One objective of the project is to improve the quality of everyday life by building and adapting technologies for a range of user groups and application domains. Consequently, it is very much concerned with developing supporting technologies based on a comprehensive understanding of user needs. Meeting this objective will require us to address fundamental and long-term research challenges in how computing technologies and concepts relate and adapt to a range of everyday domestic environments, including those characterised as "care" settings. The project employs a multidisciplinary research team to facilitate the development of enabling technologies to assist care in the community for particular user groups with different support needs. The general aim is to examine how digital technology can be used to provide various kinds of support to sheltered housing residents and their staff.

Gaining a comprehensive understanding of needs or a perspicuous view on user requirements in this domain poses a number of interesting methodological challenges; indeed this book is testament to this. It is not just that many of the important ethical and deployment issues concerning the development, deployment and evaluation of real systems remain unexplored, but that methods for eliciting needs in such a complex setting are relatively underdeveloped. Moreover, any system of determining needs must reflect the complexity of this multifaceted state (Sheaff, 1996). There are similar complex issues concerning the translation of the identified needs into a realistic, practical solution that can "enable" or support the person within their daily routine.

9.1.1 Home Environments and Social Care

Most disabled people want to live in the community as independently as possible. The extent to which that can be achieved depends to a large extent on the accessibility of the built environment, at home and in public. Few homes are built with any real thought for more complex individual needs of the people who may live or use them. When physical disability prevents convenient independent living the first option usually considered is to try and adapt the home (Bradford, 1998).

Domestic environments in general and "care" settings in particular are very different spaces from working environments and represent a very different set of challenges for those involved in the design of systems. One of these challenges centres on various conceptions of "disability". Accounts of disability from sociology and social policy have conceptualised the "problem" of disability using a range of theoretical approaches and models. The analytical focus of these approaches has shifted over time, and an account of this development would be useful here, as each has implications for what may be deemed the "appropriate" methodological stance for "design with care". This necessarily brief overview does not give the full detail of all such approaches, but indicates how the "problem" of disability has been theorised. Although differing in emphasis, many sociological accounts of disability have essentially been shaped by a Parsonian paradigm with its attendant notion of the sick role. Briefly put, these functionalist accounts expand on the notion of the sick role where the disabled (person) gives over the shaping of their lives to medical professionals. It is the responsibility of the medical professional to alleviate their "abhorrent and undesirable" situation (Parsons, 1951). However, whereas the "sick role" is a temporary one, the "impaired" or "disabled" role is one where the individual has "accepted dependency" (Oliver, 1986), particularly on the medical profession.

This "medical" or "individual" model approach further developed into the conceptualisation of the "rehabilitation role", which argues that an individual must "accept" their condition, making the most of their ability levels to achieve some sense of "normality". This process is defined and determined according to the criteria of medical professionals. However, it has been argued that this should be referred to as the individual model as the notion of a medical model places too much emphasis on the role of the doctor and under-emphasises the assumed psychological under- pinnings of this "disabled role" (Oliver, 1986). Implicit to this approach is the idea that the disabled person's situation involves some sense of loss, which led Oliver to name it the "personal tragedy theory". These "medical models" of disability have been critiqued for the way in which they view disabled people as somehow "lacking", unable to play a "full role" in society. Furthermore, medical models have implications for governmental research and policy. For example, Townsend (1975) argued that such views of the disabled resulted in them being marginalised and only ever addressed in piecemeal fashion by government policies. He saw that the extent to which governments would intervene in a welfare issue "did not bear comparison" to their willingness to help industry. Townsend's particular focus was on poverty but he also addressed trans- port, housing, education and anti-discrimination policies. Critiques such as Townsend's led to a change in analysis from the medical model to a "social" model of disability within sociology (Oliver, 1983). At the same time, groups such as the Liberation network were formed to give a voice to disabled groups.

The "social model" approach argues that the disabled are excluded from full roles by unnecessary societal barriers. Thus, a wheelchair user is disabled when a building does not have ramp access. Similarly, a deaf person is disabled if a service provider does not provide a minicom for them to access that service. People with learning difficulties are disabled when information is not given in a readily understandable format. In this view, the "problem" is not the disabled person, but the lack of appro- priate goods and services. The social model does not deny the role of the medical profession, but states that doctors themselves cannot deal with all aspects of the disabled person's life. It is also an attempt to redress what is seen as the balance of power between the two. This approach is most often stated as seeing the category of disability as a social construct, explained with reference to medical and political agenda. There is, however, a further analytical distinction here between those who take the social constructionist line and those who argue that disability is socially "created" – the difference being that the former see disability within the minds of the non-disabled and manifested in attitudes and practices based on negative assumptions, whereas the latter places more emphasis on the historical development of institutionalised discriminatory practices (see Finkelstein, 1980). Simply put, Finkelstein provides a three-stage historical model that goes from feudalism through to the present day. The phase which is seen as most excluding the disabled is the second phase from the nineteenth into the twentieth century, the period of industrialisation which excluded those who could not fit with Foucault's notion of the "disciplinary power of the factory". The third phase of Finkelstein's historical account forecasts the liberation of disabled people partly through the utilisation of technology. Such a view has the direct involvement of the disabled as implicit to research methodology.

The utility of social theory is generally based around claims to provide a clearer understanding (often an "explanation") of a situation or a problem. The social model of disability is no exception, being used by numerous researchers to enable a person centred understanding of disability. The model addresses disability from a social psychological perspective and locates the disabled person within the rhetoric of the socio-political framework in which disability is "socially constructed". However, the dilemmas faced by the social model of disability – in terms of effecting any kind of change – arise out of this methodological choice to attempt to give *explanatory* accounts of social life. Researchers set themselves up to settle explanatory questions and in so doing they are not so much involved with actually explaining anything but are more involved in questions concerned with *the form of explanation*. As Harper argues in the opening chapter, the social model addresses sociological rather than social issues producing sophisticated or credentialised stories that are regarded as professional improvements on everyday analysis. Through various renderings and master narratives the social model of disability ironicises ordinary experience, treating it as somehow partial and flawed in its ignorance of what is *really* going on. Ordinary activities are "made visible and are described from a perspective in which persons live out the lives they do, have the children they do, feel the feelings, think the thoughts, enter the relationships they do, all in order to permit the sociologist to solve his theoretical problems." Such sociological accounts of disability inevitably relate to *specifically sociological* concerns: and any claims to special insight are based upon sociological categories and concepts, often in direct contradiction to those used by people engaged in the course of their ordinary actions.

We advance an alternative approach for the analysis of disability, evidenced from extensive field research. This ethnomethodologically informed ethnographic approach seeks "to treat practical activities, practical circumstances, and practical . . . reasoning as topics of empirical study, and by paying to the most commonplace activities of daily life the attention usually accorded extraordinary events, seeks to learn about them as phenomena in their own right" (Garfinkel, 1967). While ethnomethodology has some notoriety for complaining that sociologists characteristically treat the members of society as "cultural dopes" the import of that point is rarely appreciated in that it makes the investigation of "common sense" understandings the focus of inquiry. Disability or impairment is considered in relation to how individuals practically perceive and understand it, and how it practically affects their everyday life, not in terms of some explanatory or prescriptive model. Our interest lies in understanding people's real needs, and the requirements for any technological intervention, through a consideration of details from case studies related to home technology and ubiquitous home computing. We suggest that when it comes to mundane technological intervention in the everyday lives of the disabled what is needed is an alternate position from which to understand disability, which considers disability "from within". This is not taking yet another sociological perspective upon the situation, but rather attending to the members' perspectives, replacing political rhetoric with recommendations for design.

9.2 Eliciting Design Requirements for Domestic Environments

> The home, it is contended, should be considered as more than just a physical entity (Dewsbury and Edge 2001).

The nature of the home, and the character of everyday home life is undergoing constant change in definition and as such is required to become responsive to the changing needs of people throughout their lifetime. Any design process requires the designer to consider the home from a proactive and lifetime perspective. When technology is incorporated within the home, the people who live with the technology on a day-to-day basis have tended to be overlooked (Tweed and Quigley, 2000). It is also important to recognise that the imposition of technology must be undertaken in such a way that it does not remove choice and control from the "user" (Fisk, 2001) who not only include the occupants of the living space but the support workers and others who have regular access to the home. Technology can be incorporated into the person's life, such that they come to depend on it, as Lupton and Seymour (2000) suggest:

Any human body using any form of technology may be interpreted as in some way adopting prostheses to enhance its capacities. Nearly everyone in contemporary western societies has developed a close dependency on technologies to function in everyday life, such as using spectacles to see clearly or a car to achieve greater mobility. As this suggests, the category of "disability" is not fixed, but rather is fluid and shifting, a continuum rather than a dichotomy.

The Digital Care project has begun to explore some of the methodological options open to those working in the domestic domain, in particular, the translation of research into design recommendations and the attempt to uncover, elicit or validate "requirements". The problem is that research in these contexts is often regarded as not merely difficult but often inappropriate and intrusive. The deeply personal nature of many social activities limits just *what* can be investigated, as well as *how* it can be investigated, and reporting the interactional elements in a range of activities and contexts is often difficult. These and other delicate issues represent potentially obdurate problems and methodological responses have taken a number of forms. At present the Digital Care project research method for technology development includes experimenting with combinations of ethnographic study, user-centred design and evaluation and the use of "cultural probes" with both residents and staff.

9.3 Research Methods for Design for Domestic Environments

Visions of what technology can do . . . are rarely based on any comprehensive understanding of needs (Tweed and Quigley, 2000).

Compared to work environments, where there is an almost embarrassing choice of methods of study, how and in what ways domestic environments may be best investigated for the purposes of design is largely an unknown quantity. However, in social research of late there has been a movement whereby much more interest is being shown in qualitative, and in particular ethnographic, methods of investigation; an interest reflected in many of the chapters in this book. In the field of CSCW (computer-supported cooperative work) ethnography has achieved some prominence as a contributor to the design of distributed and shared systems (see, for example, Hughes et al., 1994). Ethnography is one of the oldest methods in the social research armoury. Recent efforts to incorporate it into the system design process has had much to do with the somewhat belated realisation among system designers, that the success (and failure) of design depends upon the social context into which systems are placed. The more traditional and often cognitively based methods of requirements elicitation were seen as inadequate, or in need of supplementation, by methods better designed to bring out the socially organised character of work settings. It was also argued that such methods needed to be more attuned to gathering relevant data in "real world" environments; that is, settings in which systems were likely to be used rather than in laboratories or other artificial environments remote from contexts of actual system use.

Ethnography has gained prominence as a fieldwork method which could meet the needs of providing appropriate analyses of the socially organised character of settings as seen, understood and achieved by parties to the settings. Not only was it a method with a long pedigree in social research its emphasis on the *in situ* observation of interactions within their natural settings seemed eminently suited to bringing a social perspective to bear on system design. Moreover, it seemed well suited to an insistence that system design should pay more attention to the specificities of domains. Ethnography eschews generalisation and, instead, emphasises description over explanation by requiring the ethnographer to examine a social setting in its own terms rather than through a lens furnished by some theory. By placing the social actor's conceptions and activities as the centrepiece of the analysis, a more realistic and "real worldly" grounded portrayal of the interrelationship between activities, technologies, and organised settings could be produced and be of more help to system design's needs to be informed by a social perspective. While there is growing acceptance of the utility of ethnography, the approach has admitted problems. Of particular concern to many design practitioners is the practical problem of identifying generic design solutions from the situated and highly particularised and often-complex descriptions of social interaction provided by ethnographers (Hughes et al., 1994). Researchers are exploring a number of potential solutions to this problem, placing particular emphasis on the need to support communication and cooperation between ethnographers and designers in the process of abstraction and generalisation.

The setting for our project is a hostel and nearby and associated semi-independent living accommodation, managed by a charitable trust, for former psychiatric patients in a large town in the north of the UK. The hostel is the first step for patients leaving the psychiatric wards of local hospitals that are currently being closed down. In the hostel residents are provided with a room and are monitored by staff. Residents may then move on to the other, semi-independent living site of sheltered housing consisting of a number of flats and bedsits, prior to moving out to flats in the local area, or, if they are deemed to need further and continuing support, back to the hostel. The overall aim of these facilities is to develop independent living skills, to gradually introduce the patients back into the community and to allow them to support themselves. As a general, and important, principle any technology introduced into the setting should contribute to this goal in some way. A technology that merely completes a task for residents does little in promoting their independence but merely shifts reliance onto the technology.

As part of the project we are engaged in a long-term ethnographic, observational study of the work of the staff as well as conducting a series of informal, open-ended interviews with residents, studying interaction within natural "real world" settings, in a way which would not only minimise disruption but also provide speedy feedback to the design team. The kinds of the research questions in which we are interested include general questions about the organisation and coordination of domestic space as well as more specific issues to do with the availability and use of technologies and their affordances. However, the precise nature and value of the ethnographic input into design is controversial, especially since much of our experience comes from ethnographic investigation of the workplace. It may be that we require significant shifts in our investigative techniques as well as in our understanding of design, to consider how technology relates to domestic, specifically "care" settings and the requirement to support everyday living rather than productivity. One way in which we have attempted to increase the repertoire of available techniques is through the employment and adaption of "cultural probes". "Cultural probes" (Gaver et al., 1999), originating in the traditions of artist-designers rather than science and engineering, and deployed in a number of innovative design projects (e.g. the Presence project) may prove a way of supplementing ethnographic investigations. We use "cultural probes" (cameras, diaries, maps, dictaphones, photo-albums, postcards etc.) in the Digital Care project, as a way of uncovering information from a group that is difficult to research by other means and as a way of prompting responses to users' emotional, aesthetic, and social values and habits. The probes furthermore provide an engaging and effective way to open an interesting dialogue with users. Sensitivity to the feelings of the participants who agreed to be involved in our study involved this choice of a range of "sympathetic" data gathering techniques.

The eclectic approach adopted by this project was part of an attempt to meet some of the ethical and moral dilemmas through careful involvement and acknowledgement of users in the design process. One particular technical concern, perhaps a dominant if unusual concern for a research project, is that of dependability and associated issues of diversity, responsibility and timeliness (see the DIRC project). Given the care setting it is imperative that technologies designed for the setting are reliable and dependable. However, among the technical challenges are other "social" issues concerning the location of the interface, the generalisability of design solutions, the transfer of skills to real world situations, and support for independent living in the community. These challenges highlight some of the moral and ethical components of the design enterprise, in particular the need to carefully think through and balance issues of "empowerment" and "dependence". As Gitlin (1995) suggests, technology can present dramatic compromises in social activities, role definition and identity. Consequently, the challenge for the project is to provide support for individuals in the move towards independent living, rather than create new, technological, forms of dependence. This requires a certain ethical awareness and recognition of the various ways that technology can impinge on individual care pathways and a sensitivity towards the social implications of any technological intervention. Embodying a philosophy of care into design – in the form of encouraging a move towards independent living – necessitates considering issues of empowerment and dependence and then thinking how these might usefully become incorporated into design guidelines.

9.3.1 Supporting Awareness: Security and Medication

Our early ethnographic fieldwork indicated some major preoccupations of both residents and care workers – all of which centre on supporting various forms of "awareness" and present particular problems for the design of appropriate technology. First, there is an absolutely overwhelming and, given the circumstances, understandable preoccupation with security. Situated on the edge of a "difficult" council estate, residents and staff have been subjected to frequent physical and verbal attacks. Attacks and verbal abuse by children, for example, has resulted in the gates being locked at 4.00 p.m. each day – when the school day ends – and some residents will only travel outside the accommodation by taxi. Paradoxically, the iron railings and gates and the CCTV cameras installed for the residents' security mark them out as somehow "different" to the rest of the community and therefor the focus for possible attacks, occasionally fostered by ill-informed media "moral panics". Consequently residents are increasingly cut off from the outside community and their friends. These circumstances pose fascinating, if distressing, problems for the design of domestic technologies suggesting important connections between the home environment and the outside world. The main locations for the attacks are the road between the hostel and the semi-independent living accommodation and the park next to the accommodation leading into town. In these circumstances, a security/monitoring system that would allow staff to monitor residents travelling between sites in order to increase the sense of safety, reduce anxiety and reassure residents. Such a system may also, serendipitously, contribute to greater community awareness among both residents and staff.

In order to encourage residents to feel safer while travelling between sites, or into town, we are investigating the potential for developing personal panic alarms. When activated, such alarms would alert staff as to the identity and location of the person in distress. The alarm needs to be lightweight and should not have any significant commercial value because of fears of encouraging theft and, paradoxically, further assaults. Most importantly, the device needs to be highly dependable both in terms of location accuracy and the ability to communicate the distress call in a timely manner. The approach that we are currently considering is to deploy a device that incorporates a GPS (global positioning system) receiver and transmits the user's current coordinates via a GSM (global services mobile) connection whenever the alarm button is pressed. Tests in the area reveal that the view of satellites by the GPS receiver is very good. However, if residents did wander into an area where a GPS fix could not be obtained then this would clearly present a real problem. For these reasons, we are designing the unit in order to provide its user with simple but immediate feedback if there is any problem with obtaining a location fix and/or communicating the distress call.

9.3.2 Supporting Medication: Fieldwork

Another important concern of both residents and staff focuses on issues surrounding the routine taking of daily medication. Many of the residents are on daily medication regimes and at the initial meetings as well as in interview a number of residents expressed concern about the possible grave consequences of them forgetting to take their medication. Observation and interview confirm the role of the medication regime in the maintenance of normal everyday life and residents emphasised their often graphic fears and anxieties over the likely consequences of forgetting their medication.

Medication issues – dosage, delivery of "medi-packs", reminders, reassuring residents about delivery and so on – also feature heavily in the everyday work of the staff. At the hostel medication is kept in a locked drug cabinet, distributed by the staff when required with records kept in a written log. At the semi-independent living site patients must manage their own medication and, as stated, it is a source of continuing anxiety. Although provided with a week's supply of packaged daily doses by the pharmacy – "medi-packs" – there is some concern that they may either forget to take their medication or accidentally overdose. Technical devices that may prove useful in these circumstances are various medication reminders that help patients manage their own medication, that is, when to take it, record acknowledgements of reminders and so on, allied with a system to automate the recording of drug information. But the functionality of any technology provided must be carefully considered and sensitively deployed. The devices are intended to act as "reminders" to residents to take their medication and are not indicators that any medication has been taken and obviously such devices must be dependable as failure of the technology could have potentially disastrous consequences.

The initial studies have identified a range of requirements regarding the resident's medication and the design of technologies to support the medication regime (Cheverst et al., 2001). In the semi-independent living area residents are expected to manage their own medication and weekly supplies are provided by the pharmacy packaged into individual doses within a plastic container known as a "medi-pack". This arrangement causes anxiety and inconvenience for both staff and residents. Residents, who have previously relied on the staff to provide their medication at the correct time, must now depend on their selves to remember what to take and when, leading to worries about missed medication, taking pills at the wrong time or even accidental overdoses. This, in turn, leads to residents relying on staff to provide reassurance about the medication and in some cases reminders of when and what to take. This kind of reliance is of course, detrimental to the aims of the semi-independent unit and a solution that bridges the two stages was thought to be desirable. In order to achieve this, intermediary stage residents primarily need a system that will reassure them that they are following the correct regimen, while leaving the task of managing their medication in their own hands. It is important that the system does not take over the task for them completely as many commercial products attempt to do by fully automating the dispensing of drugs at the correct time. The aim here is not to automate a task and remove a cognitive load, but to encourage self-reliance and allay any fears of getting it wrong. In addition to these requirements it would also be desirable to provide some form of unobtrusive feedback, accessible by the staff for monitoring the residents' progress. This function of the system may also be used to alert the staff to possible problems such as a deliberate overdose.

Where residents are responsible for taking their own medication, this fact has significant implications for the way in which medication is monitored and tracked. One possibility we have explored is building certain reminder and recording features into the "medi-packs" themselves. While this will not control the medication regime to prevent deliberate overdosing, it may contribute to the prevention of accidental overdosing. Some instances from the early fieldwork – coincidentally occurring on the same day – illustrate this point. In one case the care worker, following a phone call from the resident's doctor was concerned to intercept the delivery of a "medi-pack" in order to replace one dosage of tablets with another. In another incident there was some concern that an elderly resident was accidentally overdosing as a consequence of the design and delivery system for the "medi-packs". As the "medi-packs" are delivered from the pharmacy at about 6.30 p.m. the resident was required to take only the evening dose for that day, leaving the two earlier doses to be taken the next week. Problems were arising both because the resident, used to emptying each daily dose, was accidentally overdosing by taking all the medication for the delivery day, but also was being left with no morning or afternoon medication for the same day on the following week. Finally, one of the residents deliberately overdosed by taking all the medication in the newly delivered "medi-pack". This incident also highlighted other issues to do with medication and the recording of, access to and integration of information as the care worker gave information on the resident and the medication to the ambulance service.

9.4   Design With Care: Moving Towards Appropriate Design

> The challenge for designers, then, is to pay heed to the stable and compelling routines of the home, rather than external factors, including the abilities of the technology itself. These routines are subtle, complex, and ill-articulated, if they are articulated at all . . . Only by grounding our designs in such realities of the home will we have a better chance to minimize, or at least predict, the effects of our technologies (Edwards and Grinter, 2001).

In moving towards appropriate design – for "design with care" – there is a perceptual shift that is required in order to determine the needs of the occupant(s) and reflect these needs within the overall design (Dewsbury, 2001). In "designing with care" inclusive design criteria are required before technology is even considered and a long-term view of a person's condition should be undertaken in the assessment. Undertaking a full user needs assessment is essential in order to determine if technology is appropriate to meet the needs of the person. Such assessments should consider how the person is to interact with the technology from a psychological, emotional, physical and social perspective. Clearly technology should not be seen as a panacea and while viewing technology as enabling and empowering is essential to the design process, it is important to recognise that inappropriate design is disabling, debilitating and disempowering.

While this may well be true of design in general, certainly in designing for care environments the entire design process is one of iteration (Clarkson and Keates, 2001), in which problem specification, matching the system to the real world and evaluation should be a continuous process. In this view design issues do not cease with the initial deployment of the device or initial evaluation, with everything occurring after this being given the status of "maintenance". Instead, deployment is regarded as yet another opportunity for design considerations to be highlighted, challenged and reassessed. Figure 9.1 illustrates this procedure.

These guidelines demonstrate the conflicting requirements that the designer faces and illuminate the decision-making process that produces effective and robust designs. There can never be a design that will meet the needs of all, the universal design process can only make certain considerations come to the fore.

Edwards and Grinter (2001), for example, present a number of challenges to be overcome for smart home technologies – "to produce domestic computing technology that is not simply ubiquitous, but also calm" – that stress the technical, social and ethical directions of ubiquitous computing in domestic environments. Technical concerns focus on questions of interoperability, manageability, and reliability; social and ethical concerns highlight the adoption of domestic technologies and the implications of such technologies. The ways devices are used may need

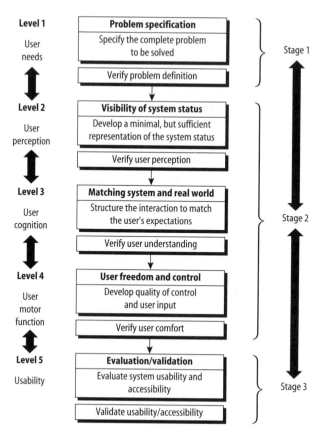

Figure 9.1 The methodological iteration of the design process (adapted from Clarkson and Keates, 2001).

to be reflected in configurations, security parameters, and device interactions. Designing for reliability requires devoting substantial time and resources that must be integrated into the development cultures; "creating a development culture that can produce reliable devices consistently ... This challenge extends beyond the research community to those who develop, deliver, regulate, and consume these new services". Understanding how technologies fit into daily routines is one aspect of design but designers also need to be aware of the broader social effects of technology. There are unforeseen and unpredictable social consequences that can arise when technology is placed into the domestic setting – and the home is perhaps an especially complex and volatile setting: "We believe that the chief challenge that will be faced by the designers (and, potentially, the occupants) . . . is balancing the desire for innovative technological capabilities with the desire for a domestic lifestyle that is easy, calming and, at least in terms of technology, predictable".

As the chapters in this book make clear, when considering design for domestic environments, the technological approaches traditionally used to determine the design and likely role of technologies need to be supplemented by detailed investigations into the everyday life and needs of domestic users. Particularly when considering users with various kinds of disability both the overall design of the home and the embedding of "intelligence" into a wide range of everyday appliances often appear to depend on particular, often unverified, models of the social and spatial organisation of the household and domestic activity. Just what is meant by design is, of course, very broad indeed. What has emerged from our ethnographic investigations even in a domestic setting as "unconventional" as community care is of everyday domestic life – such as the routine taking of medication – as sets of activities made orderly. From our perspective design is concerned with interventions into this orderliness – to supporting everyday activities in various ways by impacting on timeliness, reliability, dependability, safety or security. In this way a "philosophy of care" can be integrated into the design of domestic environments and ubiquitous computing in much the same way as other philosophies, the "scientific" and the "modern" have already been incorporated (see Banta, 1993, for example, on "washability" as an emerging design principle in the early 20th century). Accordingly, while these are difficult challenges, they should not be overestimated. Despite "hyped-up" visions of technology as a means of completely transforming home life, successful forms of domestic interactive technologies, no matter how radical, are successful precisely because they are quite routinely, "made at home" with the social organisation of the domestic environment. Uncovering the detailed nature of the social organisation of domestic life is, consequently, as this book suggests, essential to both design and development.

Acknowledgements

This work is funded by the UK Engineering and Physical Sciences Research Council, EQUATOR and Dependability (DIRC) Interdisciplinary Research Collaborations. We also thank the Croftlands Charitable Trust for their support.

References

Banta, M (1993) *Taylored Lives: Narrative Productions in the Age of Taylor, Veblen, and Ford,* Chicago: University of Chicago Press.

Bradford, I (1998) "The Adaption Process", in Bull R, *Housing Options for Disabled People*, London and Philidelphia: Jessica Kingsley.

Cheverst, K, Cobb, S, Hemmings, T, Kember, S, Friday, A, Phillips, P, Procter, R, Rodden, T and Rouncefield, M (2001) "Design with Care", EDRA32 2001 Edinburgh, July 2001, Workshop Symposium on "Adding the User into the Design of the Home in the Digital Age".

Clarkson, PJ and Keates, S (2001) "A Practical Inclusive Design Approach", *Proceedings of INCLUDE 2001*, London, pp. 72–73. Helen Hamlyn Research Centre, Royal College of Art.

Crabtree, A, Hemmings, T and Rouncefield, M (eds.) (2001) *Proceedings of the 1st Equator Workshop on Ubiquitous Computing in Domestic Environments*, Nottingham, 13–14 September 2001. Nottingham University.

Dewsbury, G (2001) "The Social and Psychological Aspects of Smart Home Technology within the Care Sector", *New Technology in Human Services*, Vol. 14, Nos. 1–2, pp. 9–17.

Dewsbury, G and Edge, M (2001) "Designing the Home to Meet the Needs of Tomorrow . . . Today: Smart Technology, Health and Well-being", *Open House International*, Vol. 26, No. 2, http://www.smartthinking. ukideas.com/_OHI.pdf

DIRC Project: http://www.dirc.org.uk/

Edwards, K and Grinter, R (2001) "At Home with Ubiquitous Computing: Seven Challenges", in GD Abowd, B Brumitt, SAN Shafer (eds.), *Proceedings of Ubicomp 2001*, LNCS 2201, pp. 256–72. Berlin Heidelberg: Springer-Verlag.

English-Luek, C and Darrah, C (1997) "The Infomated Households Project", *Practicing Anthropology*, Vol. 19. (4) 18–22.

Equator Project: http://www.equator.ac.uk/

Finkelstein, V (1980) *Attitudes and Disabled People: Issues for Discussion*, New York: World Rehabilitation Fund.

Fisk, M (2001) "In View Magazine: Focus on Technical Issues in Housing", May 2001, Northern Ireland Housing Executive, pp. 30–31.

Garfinkel, H. (1967) *Studies in Ethnomethodology*, New York. Englewood Cliffs.

Gaver, W, Dunne, A, and Pacenti, E (1999) "Design: Cultural Probes", in *Interactions: New Visions of Human-Computer Interaction*, Danvers, MA: ACM Inc.

Gitlin, L (1995) "Why Older People Accept or Reject Assistive Technology", *Generations: Journal of the American Society of Ageing*, Vol. 19, No. 1. 41–46.

Hughes, JA, King, V, Rodden, T and Andersen, H (1994) "Moving Out from the Control Room: Ethnography in System Design", in *Proceedings of CSCW '94*, Chapel Hill, North Carolina. ACM Press.

Hughes, J, O'Brien, J and Rodden, T (1998) "Understanding Technology in Domestic Environments: Lessons for Cooperative Buildings", in *Proceedings of CoBuild '98*.

Hughes, JA, O'Brien, J, Rodden, T, Rouncefield, M and Viller, S (2000) "Patterns of Home Life: Informing Design for Domestic Environments", in *Personal Technologies*, Special Issue on Domestic Personal Computing Vol. 4, pp. 25–38, London: Springer-Verlag.

Lupton, D and Seymour, W (2000) "Technology, Selfhood and Physical Disability", *Social Science and Medicine*, Vol. 50, pp. 1851–62.

Mateas, M, Salvador, T, Scholtz, J and Sorensen, D (1996) "Engineering Ethnography in the Home", in *Proceedings of CHI '96*, Vancouver, BC: ACM Press.

Moran, R (1993) "The Electronic Home: Social and Spatial Aspects", Report of the EC's European Foundation for the Improvement of Living and Working Conditions, Luxembourg: Office for Official Publications of the European Communities.

Morley, D (1986) *Family Television*, London: Comedia.

O'Brien, J and Rodden, T (1997) "Interactive Systems in Domestic Environments", in *Proceedings of DIS'97*, pp. 247–59, Amsterdam: ACM Press.

O'Brien, J, Hughes, JA, Rodden, T and Rouncefield, M (1999) "Bringing IT all Back Home: Interactive Systems and Domestic Environments", *Transactions on Computer-Human Interaction*, Vol. 6, No. 3 (June), ACM Press.

Oliver, M (1983) *Social Work with Disabled People*, Basingstoke: Macmillan.

Oliver, M (1986) "Social Policy and Disability: Some Theoretical Issues", *Disability, Handicap and Society*, Vol. 1, No. 1, pp. 5–18.

Parsons, T (1951) *The Social System*, London: Routledge & Kegan Paul.

Sheaff, R (1996) *The Need for Healthcare*, London: Routledge.

Silverstone, R (1994) *Television and Everyday Life*, London: Routledge.

Townsend, P (1975) *Sociology and Social Policy*, Harmondsworth: Penguin Education.

Tweed, C and Quigley, G (2000) *The Design and Technological Feasibility of Home Systems for the Elderly*, Belfast: The Queens University.

Venkatesh, A (1985) "A Conceptualisation of the Household/Technology Interaction", in *Advances in Consumer Research*, Vol. 12, pp. 151–55, Ann Arbour, MI: Association of Consumer Research.

Venkatesh, A (1995) "Computers and Multimedia and the American Household – The Use and Impact: Past, Present and the Future", Presentation at Research Forum, Home of the Future Meets Cyberspace, Intel Research Council, June 29, Hillsboro, Oregon.

Venkatesh, A (1996) "Computers and Other Interactive Technologies for the Home", *Communications of the ACM*, Vol. 39, No. 12, pp. 47–54.

Part 3
The Home of the Future

Towards the Unremarkable Computer: Making Technology at Home in Domestic Routine

Peter Tolmie, James Pycock, Tim Diggins, Allan MacLean
and Alain Karsenty

10.1 Introduction

In this chapter we take a look at some issues surrounding the notion of "ubiquitous computing", and in particular we consider how ubiquitous computing might ever come to have the kind of character that would enable it to support things like domestic routines. Put simply, when we have more computers everywhere around us, how are we going to interact with them all? More importantly, perhaps, how are we going to avoid having to interact with them in the way that we currently have to interact with desktop computers? How are we going to use computational power in ways that will, in Mark Weiser's words, make it truly "invisible in use" (Weiser, 1994b)?

Furthermore, if we believe that computing is going to become truly ubiquitous, then we also have to believe that it is going to be *everywhere*. This means it will become a part of every aspect of our lives not just our work and certainly not just on our desk at work. Yet it would seem that "the office" is almost *the* sole focus for ubiquitous computing. Even when Mark Weiser first articulated the notion of ubiquitous computing the office was the default domain:

> Inspired by the social scientists, philosophers, and anthropologists at PARC, we have been trying to take a radical look at what computing and networking ought to be like. We believe that people live through their practices and tacit knowledge so that the most powerful things are those that are effectively invisible in use. This is a challenge that affects all of computer science. Our preliminary approach: Activate the world. Provide hundreds of wireless computing devices per person per office . . . (Weiser, 1994a)

And it has largely stayed like that ever since.

In our research, however, we have been considering the notion of ubiquitous computing in the context of another domain – the home. As a number of the other authors in this book and beyond (e.g. Venkatesh,

1996) point out, the home is a place where technology is already changing: there are white goods with more embedded computation; there is interactive television, broadband, wireless networks; and so on. There are also social changes that are impacting the home with more people working at home and finding themselves answerable to "any time any place" demands that put pressure on the boundaries between home and work. So if there was ever a place that could prove to be radically demanding, a place that would set some of the greatest challenges to the acceptance of computing in our lives, then it is within our *own* homes.

However, there is an increasingly obvious disparity between the traditions of technology design for the office and the traditions of technology design for the home. While the vocabulary of office technology has revolved around tasks, processes, functionality, productivity etc., as Randall and others note elsewhere in this volume, design for the home has been more concerned with dimensions such as lifestyle, aspirations, emotions, aesthetics, and so on. Yet, as ubiquitous computing takes hold, we can expect that computing will increasingly expand from the work domain and will become embedded within home appliances and domestic environments, setting these two technology and design traditions on a potential collision course.

Additionally, we have been motivated by a belief that, if we look at ubiquity in the domestic domain where radical differences between the home and the office may become manifest, it may well oblige us to re-evaluate many of the assumptions buried within prevalent views of ubiquitous computing. Alternative domains have a habit of challenging consensus and questioning engrained perspectives in that way.

Our overall goal was to understand how ubiquitous computing might arrive and make its place in domestic life. The strategy we adopted in examining home environments was first and foremost to "let them speak for themselves". By bringing to bear an approach adopted by a number of the other authors in this volume, known as "ethnomethodologically informed ethnography", which, as both Cheverst et al. and Randall point out in their chapters, is a strong feature of recent work in human-computer interaction (HCI) and computer-supported cooperative work (CSCW), we sought to arrive at a thoroughly empirically grounded and pre-theoretical understanding of the domain. This approach involves *in situ* and *in vivo* observation, where the ethnographer seeks to become not just a passive observer, but a competent member in some setting, thereby gaining access to members' relevances and understandings. This approach carries a strong injunction to avoid viewing the setting through some pre-given theoretical lens. The ultimate aim here is to uncover both the actual lived details of phenomena and to bring out the *ethno-methods* (Garfinkel, 1967) and tacit resources whereby things come to look the way they do. In this particular project one of us participated in the domestic lives of five households over the course of a year, typically spending several weeks with each.

10.1.1 The Glue of Domestic Life

Although our starting point was simply a general interest in domestic environments and ubiquitous computing, as we set about looking at the everyday phenomena of life and work within the home, one of the things that struck us most forcefully was the prevalence of routines and how much turned upon them. We had not set out to analyse domestic routines but it became evident that they were highly significant in home life and had intriguing characteristics. Indeed, in some of the homes we studied where work was also being done, work routines were typically made subservient to domestic routines. Work was seen as a thing that (within certain confines) could be done anytime within the day while breakfast had to be *now*, the children had to get to school *now*, and so on.

There is a sense in which routines are the very glue of everyday life, encompassing innumerable things we take for granted such that each ordinary enterprise can be undertaken unhesitatingly. In his chapter Randall comments on how fleetingly families are "collective entities". The fact of the matter is that in the home highly disparate priorities of different family members have to be coordinated regularly without the commonality of an orientation to some shared work objective to bind them together. Routines, then, help provide the grounds whereby the business of home life gets done. Routines mean that people can get out the door, feed themselves, put the children to bed, and so on, without having to eternally take pause and invent sequences of action anew, or open up their every facet for inspection or challenge, or to constantly have to *account* for what they are doing with explanations or rationales.

Routines, then, can be seen to have a tremendous significance in this domain. Furthermore, domestic technology is both implicated in household routines and influenced by them, something that is also commented upon by Frohlich and Kraut elsewhere in this volume. Maintaining the equilibrium of domestic life frequently involves the routine use of technology. This use of technology both facilitates and reflects the mundaneity of daily household routines. As O'Brien and Rodden note:

> sophisticated sets of understandings exist as to the nature of the routines and preferences of others within the house, who not to disturb when, who is claiming a particular piece of the household space by watching the programme that they always watch and the like (O'Brien and Rodden, 1997).

However, it is not a simple matter of household routines changing to accommodate each new piece of technology. The work by O'Brien and Rodden and other more recent studies of domestic technology use (O'Brien et al., 1999; Hughes et al., 2000; Crabtree et al., 2002; and Tolmie et al., 2002), not to mention both Frohlich and Kraut and Cheverst et al. in this volume, have all tended to confirm an observation that was first offered by Harvey Sacks in the early 1970s with regard to the use of the telephone:

> This technical apparatus is, then, being made at home with the rest of our world. And that's a thing that's routinely being done, and it's the source for the failures of technocratic dreams that if only we introduced some fantastic new communication machine the world will be transformed. Where what happens is that the object is made at home in the world that has whatever organization it already has (Sacks, 1992, pp. 548–49).

It is clear, then, that in environments where rationales of productivity and efficiency are not at the fore, and where aesthetic configuration is often given considerably greater emphasis (O'Brien and Rodden, 1997), how technology looks, where it is placed, how "visible" or "invisible" it is, how to hand it is, and how much work it takes to make it work (Bowers, 1994), are all clearly significant. Furthermore, they are unlikely to be significant in the same ways as they are for workplace technologies. But what does it take for a computer to "disappear"? And how do technologies get "made at home"?

10.2 Work Routines

Despite O'Brien and Rodden's (1997) discussion of routines in relation to the "equilibrium" of the home, there is little empirical understanding of the fundamental nature of routines in domestic life to date, something that both Randall and Cheverst et al. also pass comment upon in their chapters. Additionally, while O'Brien and Rodden (1997) (and also Venkatesh, 1996) make some tentative suggestions for the design of domestic technologies, no means have yet been found to allow for an understanding of domestic routines to impact upon the design of domestic technologies in a way which is comparable to the impact that the study of routines in the *office* environment has had on fields such as CSCW.

The significance of the notion of "routines" came to the fore in the late 1970s and early 1980s when technology developers began to explore ideas of "office automation" (see for example Ellis and Nutl, 1980; Zisman, 1977). However, it was the field studies of researchers like Wynn (1979) and Suchman (1983) that first demonstrated the rich and complex nature of allegedly repetitive activities and the skilled and cooperative decision-making and negotiation necessary to "get the work done". Suchman (1987) in particular was able to suggest a radically different sense to "routine", illustrating the importance for design of taking an ethnographic orientation to the status of procedural plans by seeing them as accomplished products rather than as structures which stand behind the work.

Embedding representations of routines within systems (such as workflow tools) was seen to change the status of those representations from being a resource for situated action to becoming something to be merely enacted programmatically. A focus upon supporting work with resources rather than automating representations of routines has now become a distinctive characteristic of CSCW where "routineness" is recognised to be an accomplishment produced through the practised exercise of complex skills.

10.3 Returning to the Home

We would certainly not wish to understate the significance of the above body of research. Indeed, the work of Suchman was motivated by the same core interest and approach as our own. However, in CSCW research this is now a well-worn path, where the primary focus has remained upon *work* practices and typically the *office*. While Randall has sought to extend CSCW approaches to domestic environments elsewhere in this volume, we have taken a somewhat different approach by setting them aside for the time being in order to take a fresh look at routines and to treat the home as a substantively new domain. We take no position on the relative validity of these perspectives and we feel that both have much to offer. While Randall seeks to understand the similarities and differences between work and domestic environments, we have simply chosen to set aside for the time being existing paradigms and, so to speak, discover the home anew, something that echoes closely Hindus's (1999) injunction not to generalise understandings of the workplace to the home.

It should be said at this point that domestic routines cover a wide range of phenomena with many research implications. Our aim in this chapter is therefore a modest one: fully endorsing the view expressed by Cheverst et al. in their chapter that routines in the household are socially organised through and through, we seek to elaborate upon this by beginning to identify, through empirical materials, some of the features of things that have a *routine character* in the home. This point requires emphasis. Our intention here is not only to highlight the centrality of routines to domestic life, though that is certainly the case and domestic routines are a neglected topic of investigation in the literature. Rather it is the *character* of these routine activities which we want to explore. As we shall shortly demonstrate, routines have some interesting characteristics: they appear to be undertaken without hesitation; people do not have to invent each new instance of a routine as if it were a new occasion; and people can undertake routines without having to account for what it is that they are doing by providing some form of explanation or rationale. In short, routines are things that can be just "got on with". They may well be complex but, nevertheless, one of their most powerful features is that they are taken for granted.

We have not, in that case, set out to produce a list or taxonomy of routines or a measure of their generality. Our aim has not even been simply to compare domestic and office routines. Rather, we have explicitly chosen to look at examples of things that we might call routine in order to attempt to understand, *foundationally*, what it is that gives these courses of action a routine character.

10.4 Routine Instances

10.4.1 The Knock on the Door

Our first instances of interest are two distinct but related observations of the domestic round of a family with two children, one aged 12 and the other aged 9, collected on different days. Both of the instances occur at the time the mother, whom we shall call Christine, departs to pick up her youngest daughter, whom we call Susie, from school. They also involve the neighbour (and sister-in-law) whom we call Louise and who has a child at the same school.

Instance 1a

Christine was sitting at the end of the garden in the sunshine drinking a cup of tea. It is 3.00 p.m. and she is heading back to the house to get ready to fetch Susie from school. She goes into the kitchen through the back door, shuts and locks it and closes the kitchen window, before putting away some shopping that she has left out, picking up her mobile phone and going through into the hall. She puts a few items on the stairs and goes into the living room. There is a knock at the door. She goes into the hall and half opens the door and, without looking to see who is at the door or giving any verbal response, goes back into the living room to finish what she is doing. Then she goes out onto the street, shutting the door behind her. Her next-door neighbour, Louise, is already walking slowly up the street and looks to Christine as she comes out. Christine heads over to Louise, commenting on the heat, and they walk up the road together towards the school.

Instance 1b

On another day, it is a couple of minutes past 3.00 p.m. Christine has just gone into the house from the back garden and has been going round closing doors and windows. A moment later the door to both her house and Louise's house next-door, open and they come out down their respective paths. They look at one another and Christine says, "That was good timing". Louise pauses at the end of her path and when Christine reaches her they walk off up the road together in the direction of the school.

As some additional background, it is worth noting that Christine and Louise have never discussed this arrangement, it having "just evolved". Finding they were leaving at the same time, they had started to walk to the school together, with whoever comes out first knocking on the other's door before heading off. Neither of them waits if the other one does not come out.

10.4.2 The "Message" in the Knocking

We might first of all wonder about what is accomplished through this knock on the door. Actions such as "knocking on a door" can achieve various things beyond just making a sound on a surface. Things can be "done in the doing" of a knock – such as a statement that "I'm here" or a means to "check for absence prior to entry" or a confirmation of the ownership of a space and the rights of access to it. Clearly a knock such as this could be a "summons". However, an ordinary thing about a summons is that the summoner waits for the summoned to answer, yet that is clearly not what is going on here. In instance 1a, Louise knocks on the door and then walks away without waiting for Christine to appear. This is not, however, some form of peculiar game. In fact, Christine in no way holds Louise accountable for that behaviour. The knock, then, is oriented to as not so much a summons as a *message*, the import of which is only locally intelligible. That is, for each of the mothers involved, the knock is *just enough* to tell them that the other mother is about to walk to the school.

10.4.3 The "Message" in Opening the Door

Another otherwise strange feature of instance 1a is the way Christine only half opens her front door and immediately returns to what she is doing without speaking to the person knocking at the door. One would typically expect that either a caller would be greeted immediately or that a half opening of the door followed by walking away would be highly accountable, prompting an apology or explanation (for instance by saying "sorry, I was just in the middle of something"). Christine, however, clearly has a solid *expectation of the implicativeness* of this knock such that she can disregard the possibility that her actions might cause offence or be held accountable. The routine has become honed such that the most minimal of actions has a wealth of significance and well-understood mutual accountabilities. In this way, Christine's half-opening of the door is *just enough* to suffice as an acknowledgement while she is involved in doing something else. The opening of the door, then, also serves as a message, whereby an announcement of imminent departure can be minimally acknowledged.

10.4.4 Situated Meanings

We now want to move on to considering how it would have been had the knock on the door taken place at some other time of day, somewhere else, or at 3.00 p.m. on a Saturday. Clearly the phenomenon here involves preparations to collect a child from school and is only intelligible at a very specific time of day, and only on certain days for certain weeks of the year. Both Christine and Louise are able to mutually orient to that local and highly precise intelligibility in such a way as to enable the coordination of one specific commonality of routines between two families. The particulars of how these sequences of actions are realised serve as resources for achieving an effectively coordinated shared routine. Central to this shared routine is that neither of the mothers "open up" the operation of it for remark or problematise its unique features (which, in relation to all the many things that knocking on a door and opening a door might amount to, are quite distinctive). In instance 1b, for example, what is remarked upon is not the practice itself but rather the perfection of this particular realisation. The beauty of instance 1b is that, in that one moment where they walk out of the door together, the very need for there to be the originally observed phenomenon, a knock on the door, simply fades away and reveals that *this is never simply about knocking on a door at all. That is only ever a resource to bring about what they are really after,* which is to walk to the school together, rather than separately and alone. A knock on the door provides for all of those occasions when they fail to walk out of their front doors at the same time as one another. But when they do, to still knock on one another's doors would be patently absurd.

This realisation of this routine relies upon the mutual intelligibility of certain very specific courses of action, *situated* courses of action that in just about any other set of circumstances might be meaningful in totally different ways. There is also a highly nuanced adaptation of wholly mundane physical and interactional resources such as knocks on doors, and openings of doors. The result is that some, at first sight strange, happenings at 3.00 p.m. on a school day can add up to something meaningful yet evidently unremarkable for two mothers from different houses who want to walk together to school. So, to summarise what we have discovered here: first, specific meanings can accrue to certain activities such that they can serve to facilitate the coordination of routines (including routines across households). Secondly, these meanings can be highly particular and only locally intelligible. Thirdly, the shared understandings of the meanings are such that those doing them do not have to account for what it is that they are doing or why. Finally, these activities can be "just enough" to achieve what needs to be done and it is what is "done in the doing" (such as giving a message to notify of imminent departure) that is the matter of significance.

10.5 The Alarm Clock

Our second instance of interest is an extract from a study of a freelance language translator working at home. The translator in question, whom we shall call Lucie, lived in a small three-bedroomed house with her two children, a boy aged 12 and a girl aged 10. The previous year she had moved from doing translation work in an agency to "going it alone" at home and had converted one corner of her living room into an office. This form of translation work is paid by the word and so Lucie frequently started work early in the morning before her children had got up in order to get as much as she could done without interruption. This instance is drawn from observations of one such early morning session. Lucie has been sat at her desk since about 6.00 a.m. translating from English into French a text describing a new dieting aid. To begin with her children are asleep upstairs but over the course of this instance their morning routine begins.

> Instance 2
>
> Lucie flicks through some printed sheets on her desk and comments on how the table of contents doesn't match the text. She returns to the electronic document and continues to translate the next title, saying out loud a segment. It is 7.00 a.m. and an alarm goes off upstairs, which she shows no reaction to and continues to key in as before. When she has completed that section of text she switches her monitor off and says "it's been an hour". She pushes in the leaf to her desk, stretches, then leans on the ledge under her monitor resting on her elbows, her hands to her cheeks, drinking coffee. Once she has finished her coffee she goes into the hall to call upstairs to the children: "Bonjour mon gros doudou, Bonjour mon lapin . . .".

10.5.1 Treating as Unremarkable

Frohlich and Kraut in their chapter point to how people manifestly display less attention to tasks that have become "habitual". We feel that it is important not to underplay or misconstrue this point and we would like to take it somewhat further in a rather different, though not necessarily uncomplementary direction. One feature we would particularly like to draw attention to here is the way Lucie manifestly *ignores* the alarm going off upstairs at 7.00 a.m., despite going to the foot of the stairs to call up to the children a short while later. To the ethnographer sitting beside her the alarm going off is a notable enough event for it to be recorded in the field notes. Despite the fact that Lucie regularly commented to the ethnographer about numerous other events, here the alarm passes by without remark.

Having reported upon what Lucie actually did, let us for a moment consider what she did *not* do, what other plausible actions did not happen. For example, when the alarm went off she did not draw attention to it by saying, for example, "whatever is that?". To have done so would have *marked out* the *"unusualness"* of the occurrence, perhaps prompting investigation or the seeking out of some explanatory account (for instance "the alarm has been set wrong somehow"). Similarly, she did not comment "there goes the alarm again" which suggests that the alarm is a regular but still *"notable"* occurrence. Alternatively she might perhaps have said something like "Oh, is it that time already?", through which comment she would not be marking out the alarm per se but rather the alarm would be the thing which prompted her to notice the passing of time, just as other things can prompt such a thought. This would not be an account for the alarm going off but a remark about something else.

She could, then, have commented in many ways and in doing so could have suggested many things such as marking out how unusual the alarm was, how regularly it interrupts her, what an irritation it is, what it has made her think of and so forth. However, none of these happen but instead what happens is that she in no way, shape or form marks out the going off of the alarm – not a twitch, not a blink, not a sigh. If she had commented upon it that would have made it a different phenomena, in that through Lucie's total lack of reaction to the alarm she displays her orientation to it as something wholly *unremarkable*. By *manifestly* not marking this out she provides for the sense of the going off of the alarm upstairs at 7.00 a.m. as being a matter of routine, for *who would comment upon a feature of their routine as though it were somehow special?*

Furthermore, this is something she is *able* to do. That she can choose to not mark out the alarm and to *treat it as* something unremarkable makes it evident that there is then nothing inherent in the going off of the alarm that obliges her to treat it as a notable or remarkable event. The alarm is *unremarkable*.

This is not to say that people never notice elements in their routines. One can carefully watch that a pan of water does not boil over when cooking without provoking remark because it is appropriate to do that within the routine. In this way something can require concentration or careful attention but *as part of* the routine, in a manifestly unremarkable and evidently appropriate way. In this way it is *already* intelligible *in terms of* the routine and needs no further account. However the "routine" character of events *is* fundamentally undermined when to pay manifest attention to them prompts some kind of special account for that attention. To mark something out is in many ways then the exact opposite of something having a routine character and to mark out something that is normally routine has the consequence of generating a requirement to produce an account, explanation or rationale.

So we can note here that elements of routines (understandings, practices, artefacts, courses of action, etc.) achieve the status of becoming unremarkable by virtue of having been *made* routine. Consequently, they can be apparently unnoticed. Additionally, where they are obviously paid attention to this is either (1) evidently appropriate in terms of the routine and hence equally unremarkable or (2) it is remarkable and, for those engaged in a routine to remark upon the routine itself, is an accountable action.

10.5.2 The Unremarkable as a Resource

However, although the alarm going off has the status of something unremarkable, that is not to say it is a thing without import. For a start, it is a thing of import for her children. Its very mutual availability to Lucie and themselves makes not acting upon it highly accountable. In this way, it is *used*. It is a resource. It can, for example, serve to initiate other features of the everyday morning routine, such as getting out of bed, going to the bathroom, getting dressed and so forth. That it is used is revealed by Lucie's subsequent movement to the foot of the stairs to call up to the children. This also suggests an orientation to the alarm as something "nodal", a thing upon which many other things may turn. So not remarking upon the alarm going off is certainly not dismissal.

Similarly, though aspects of routines may never be directly remarked upon, not responding to their implicativeness is accountable, and accountable in the very terms of what is usually unremarked. For example not getting up in the morning might prompt a remark such as "didn't you hear the alarm going off?". So in the example we have described it is not the case that she has not noticed or is not attending to the alarm going off. *Rather she is not marking out through some visible display that this is notable because to display that would be to make her accountable for her interest in its significance.*

Finally, one can imagine instances where she might display some interest in the alarm *not* going off (perhaps by noticing the time and realising the alarm has failed). Should an alarm fail to go off that failure could itself be quite specifically marked out. Alarms, then, can be perceptually visible yet practically invisible *in use*, as part of what has been made routine. Relatedly, they can be perceptually non-existent (through, for example, failure to go off) yet practically marked out. What matters about the alarm here is not so much its perceptual character as its significance, a significance that can be made explicit should the alarm ever fail.

To cut to the chase, let us say that one of the prime ways in which things may be "lost to view" and "made at home" is through the orientation people adopt to them as *unremarkable*. That is, it is less a perceptual matter (though we wouldn't want to utterly deny the importance of that in some respects) and more a matter of *orientation*. In everyday life there are innumerable things that we engage with, that we do, that *other people do*, that we never trouble to concern ourselves with or pass comment upon. Everyone just gets on with doing them as though they were the most "natural" thing in the world. That is, in the course of whatever they are doing, people find these things *naturally accountable* (Garfinkel, 1967) – they require no special account for why things are done that way or look that way and we would never normally think to provide some account for them. And it is in just such circumstances that things are most "invisible", "unnoticed", "ignored", "not attended to", or whatever, and most "made at home".

So in our observations of Lucie ignoring the alarm clock going off we have discovered some further orderly features of what a "routine character" might consist of. Once again we can see that an orderly aspect of things with a routine character is that they can serve as resources for the mutual coordination of unremarkable activities (in this case, the activities of getting up and, in the previous instance, the activities of setting out on a task together). These resources are mutually available and mutually accountable for those involved in the routine. Things do of course go wrong in domestic life, alarms can fail – but failure, in contrast to accomplishment, *is* remarkable and the elements held to account when part of a routine fails are the very ones that are unremarkable at other times. Evidently not marking out an element of a routine is not equivalent to not noticing that element. In this way, artefacts that are implicated in routines can be perceptually available yet practically invisible in use. And, finally, a feature within many routines is that there are nodal occurrences that are implicative for things that follow.

10.5.3 Going to the Coffee Shop

In our final instance we seek to both delineate what we have said about what provides for some course of action having a routine character, but also to begin to demonstrate how "knowing other people's routines" can itself be a powerful resource for articulating and meshing together highly distinct orientations and goals, where it may be that one of the inter-actants is *never normally part of that routine at all.*

The instance is taken from an ethnographic study of the work of a freelance website designer and graphic artist, whom we shall call Michael, who works at home. Michael likes to focus his business upon the local community and engages with many of his clients face to face at a local coffee shop. This particular sequence of events was prompted by Michael working through a "To Do" list he keeps on his desktop in MS Word, which he checks through at the beginning of each working day.

Instance 3

Michael is greying out things he's done on his To Do list – He says about needing to do something about "John's" opening times – [John is the proprietor of a local Farm Produce Shop] – He knows John wants them changed on his poster but doesn't recall for sure what to. Michael goes to a folder on his PC titled "Posters" and clicks on a document called Farm Shop, which opens in Illustrator. Leaving the poster open he goes to phone John. However, John doesn't answer. He notes that the time is about a quarter to ten – He says he thinks he will go to the coffee shop [a small coffee house just around the corner] where he thinks he'll catch John because John usually goes in there for a coffee before opening up the Farm Shop at ten o'clock – When he gets to the coffee shop he sees John waiting at the counter – He goes up to talk to him and says about the poster, checking what times John wants to go on it. While Michael queues they talk about John's website and some advertising he wants done for some chocolate products he's going to be selling. Just before ten o'clock John goes off to open the Farm Shop and Michael says he'll call in to see him later and talk about things in more detail.

Now, so far we have looked at examples of routines that are oriented to as resources for activities within a particular household, and across two households with certain common interests. However, this instance is quite distinct in a number of ways. There is no matter-of-course requirement upon Michael that he should specifically coordinate his routine with John's and he has no particular accountability placed upon him that he should attend to John's routine at all. In direct contradistinction to our previous observations Michael quite specifically *marks out* what he knows of John's routine for comment – he knows that John goes to the coffee shop every morning before he goes to open up his own shop. Here John's activities have been made a matter of note for Michael in a way that John himself might not ordinarily take note of them. John would be unlikely to mention to, say, his family before leaving the house that he was going to the coffee shop if that was a thing he did every day because the mentioning would invite that it be seen as something out of the ordinary and specifically *significant*. John might make mention of his morning coffee as a thing he did by habit to facilitate someone like a visitor finding him, but such a *mentioning* is, importantly, a quite separate occasion to actually *going to* the coffee shop as a matter of routine.

All of these observations are not independent of one another but are, in fact, quite tightly related. It is exactly because Michael is not a member of the cohort involved in John's routine that an element of John's routine can be, for Michael, a matter of comment. Thus Michael is not accountable for having made something notable and significant out of what, for members of John's family, is necessarily taken for granted. Furthermore, in this specific instance Michael is not engaged in routine activities himself; on the contrary, his actions are specifically occasioned (by not being able to complete a "to do" item and not being able to speak to John on the telephone) and thus Michael has an explicit interest in marking out an element of John's routines which he has knowledge of. So here we have someone who is not pursuing a routine of their own but is using what they have directly noted about someone else's routine as a resource to accomplish a particular course of action. This use of other people's routines as a resource for tailoring specific actions has been noted in a number of other studies, including studies of domestic telephone use where some pretty fine-tuned judgements can get made about "when it's a good time to phone" (Lacohée and Anderson, 2001).

10.5.4 Practical Actions and Descriptive Accounts

What we are not saying, however, is that people are somehow oblivious to their routines just because they never remark upon them in the actual course of doing them. On the contrary, as Frohlich and Kraut amply demonstrate in their chapter, one can perfectly well provide a description

of a routine and justify it in the context of other activities like being interviewed. Here, for instance, we have an excerpt from an interview where a
family is describing their morning routine in the context of a study of
domestic technology use (O'Brien and Rodden, 1997):

C1:   . . . have TV on after we've finished breakfast
F: . . . after breakfast, yeah. They'll come in and she watches UK Gold
C1:   Neighbours
C2:   . . . Watching old neighbours
F: Yep, she comes in and watches it after she's had her breakfast
C1:   I miss it
F: She usually misses it because she's faffing about in the bathroom
C2:   . . . She'll sit down for five minutes while it finishes
F: In fact as soon as it finishes, we get up and put coats on . . . we know it's
      time to go to work then! (laugh)

In these cases, though, giving a description of a routine is specifically
occasioned – being asked, for example, is the motivation to answer and
the context in which you are asked guides what answer is appropriate.
The occasion that prompts the account also prompts the picking out of
details of a routine and imbues those things with certain significances.
Importantly then, *an occasioned account of a routine is different from
the actual realisation of a routine* where, to give something marked significance, is wholly contrary to just taking things for granted. Indeed things
that are taken for granted form the very *background* against which one
might take note of and mark out other activities, activities that *are* significant, relevant, distinctive or notable and are so according to the occasion
that is prompting the description.

So we can note here that there are circumstances for explicitly
remarking upon both one's own and other people's routines, but, importantly, these remarks are situatedly occasioned. One of the ways in which
people's routines become discoverable to others is through such circumstances where people explicitly provide details of their routines within
occasioned accounts. Here, for instance, is the continuation of the above
excerpt:

M: . . . I didn't know all this did I?! Eh?!
(all laugh)
M: I thought you were . . . busy doing summat!

However, another important way of discovering other people's routines
is made manifest in the case of Michael's visit to the coffee shop. Here
the availability of John's routine for Michael's inspection was a matter of
Michael's own *noticing*. He had discovered it through his own recurrent
visits to the same coffee shop in the pursuit of his own routine.

So, to summarise, people can provide accounts of their own routines
and people can be interested in the routines of others. Providing an

account of a routine, however, is occasioned and what is described as relevant within the routine is bound up with that occasion. In addition, there are appropriate motives for displaying interest in someone else's routines and such interests are also specifically occasioned (e.g. by needing to talk to them). Knowing the routines of others can serve as a resource for an activity and the routines of others can be discovered through occasioned accounts and through noticing.

10.6 Ubiquitous Computing and the Quest for the "Invisible"

We have pointed to a number of features of things that have a routine character and the strong sense in which routines are deeply unremarkable. It seems that they offer courses of *action* that are invisible in use for those who are involved in them. Returning to the agenda set by Mark Weiser (but of course developed by many others since) we would now like to consider whether we could learn from this ways to develop forms of *interaction*, which are in their own way invisible in use, in their own way unremarkable. Can we begin to address the ideal of ubiquitous computing expressed by Mark Weiser when he said that:

> For thirty years most interface design, and most computer design, has been headed down the path of the "dramatic" machine. Its highest ideal is to make a computer so exciting, so wonderful, so interesting, that we never want to be without it. A less-traveled path I call the "invisible"; its highest ideal is to make a computer so imbedded, so fitting, so natural, that we use it without even thinking about it (Weiser, 1994b).

Things with a routine character would seem to have a number of the qualities we are aiming for. They are tacit and calm rather than dramatic. They do not demand attention except when needed. They are seen but unremarked. They are used as resources for action and yet they themselves also turn upon everyday resources – doors, alarms, coffee shops etc – in ways that have a wealth of significance but in ways that have been made unremarkable. There is then much to be learnt from the routine character of these courses of action. However, the important question is whether we can turn these observations into guidelines for technology design, something that was well recognised by Edwards and Grinter (2001) in their similar emphasis upon the potential importance of, on the one hand routines, and on the other "calm technology", for the design of ubiquitous computing. This is a challenging enterprise and we do not claim to have made as much progress as we might have wished for. Nonetheless, when we look at current approaches being proposed for developing ubiquitous computing that is "invisible and embedded" we find three areas where we believe some reconsideration and perhaps some redirection is needed.

Having discussed these we will turn to a final, and highly tentative fourth point, which looks to the future and how it might be if to interact with ubiquitous computing was a part of our everyday routines.

10.6.1 Inherently Invisible, or Simply Unremarkable?

Let us start out with two photographs (Figure 10.1) from a fairly well-known research project on ambient computing (Philips Research, 2000) which show one of the current approaches being proposed for developing ubiquitous computing that is "invisible and embedded". The images are intended to show a contrast between, on the left, our current world and, on the right, the future world of ambient computing in which "All sorts of computing devices will disappear into the background of our everyday lives".

Clearly these are attempts to show to a general audience what might be meant by the very idea of "disappearing computers" and we do not mean to single out this project for criticism, nor are we unsympathetic to the intent here. However, these types of images tend to suggest a focus upon the *perceptual* visibility or invisibility of computing technologies. However, we have seen in the preceding examples that perceptual invisibility is not necessarily the same as the achievement of invisibility in use. The alarm clock example described in instance 2 involves a perceptually demanding device yet one that has been made routine. The alarm is not smaller or quieter or somehow perceptually ambient but rather, as a function of use, its *significance* has been made unremarkable. In contrast, an alarm making no sound at all could be an event that is quite specifically remarked upon.

The notion of "invisibility in use" is a difficult idea. Its full implications for the design of technology have not yet been discovered. Often "invisible in use" is understood as meaning literally (perceptually) invisible as enabled by the miniaturisation of computational technology that allows devices to become smaller and (perhaps) perceptually less visible.

a   b

Figure 10.1 "All sorts of computing devices will disappear into the background of our everyday lives" (reproduced with permission of Philips Research).

However, we believe that the design goal that was originally envisaged as part of the ubiquitous computing programme requires a different understanding (though one which may not have been helped by early examples, such as "The Dangling String" which can be read as concerning perceptual psychology of "peripheral sensory processing" (Weiser and Brown, 1996), rather than issues of a resources in action). Clearly there *are* perceptual qualities that may be involved in creating an "invisible-in-use" phenomenon (an alarm is no use unless you can hear it). Yet we feel that too narrow a focus has emerged upon the perceptual qualities of a device rather than upon how people embed these perceptual resources into routines such that they are unremarkable in use. We feel this sense of "invisibility in use" is already prefigured in the attempt to turn attention away from the search for better "inherent qualities" of computers. What is sought is not a computer that is just more intimate (Weiser, 1988) or even more intelligent (Weiser, 1994a) but rather an altogether *unremarkable computer*: "Whereas the intimate computer does your bidding, the ubiquitous computer leaves you feeling as though you did it yourself" (Weiser, 1994b). Similarly, inherent perceptual qualities regarding visibility are not the same as invisible in use. Computers that have visually disappeared, or that produce perceptually "softer" notifications are not necessarily any less present. The aim is not for a hidden computer. Indeed a computer that behaved as computers currently do and required the same form of interaction but which could not be seen or heard could be *more* remarkable, *more* present than before. The challenge for design is to go beyond simply focusing upon the perceptual qualities of devices and to make computational resources that can be unremarkably embedded into routines and that might serve to augment the courses of *action* within which people find them intelligible.

10.6.2 Augmenting Action

Here, then, we wish to move on to our second point derived from our studies of domestic routines – that it is actions that need augmenting not artefacts per se. Artefacts may need augmenting in order to augment actions, of course, but those artefacts are to be in service of the actions and their augmentation should be motivated by their role in those actions.

In fieldwork instance 1a, it is clear that everyday artefacts and actions are being used. The doors are offering hard surfaces which hands can knock on to make sounds, they are offering solid barriers which can be opened to allow entry, closed to prevent it, or opened to varying degrees. These are everyday features of the tangible world that are being manipulated using mundane competencies people have for touching and moving surfaces. However, it is also clear that much of the significance of the use of these doors comes from what is done in the doing of actions with them. The knock on the door is not only the action of lifting one's hand and connecting it to the door artefact so as to make a sound audible to those on the other side of the door. Here it is also a means to coordinate actions and make others aware that you are ready to begin a routine. These are the *significances* of these actions. Furthermore, it is apparent (as in instance 1b where both parties leave their houses together at the same time) that there can be occasions when the artefacts themselves may not need to be used at all and the aims of particular courses of action can be achieved through other means.

This suggests to us that some caution is required when considering an approach to ubiquitous computing that is based upon augmenting tangible artefacts. Again we must stress that we fully support the intent underlying the tangible interface paradigm. By attempting to make computing "so embedded, so fitting, so natural", augmenting physical artefacts becomes highly appealing (especially if these provide visible interaction mechanisms for perceptually invisible computer hardware). Furthermore, the tangible interface approach is a perfectly coherent HCI approach. Manipulating physical objects is one of people's everyday competencies and more generally available than, say, abstract computer commands and software applications. There is a logic behind developing tangible interaction mechanisms just as there has been a logic behind designing other such everyday competence-based interaction paradigms: spatially based systems (like rooms – Henderson and Card, 1986 – or virtual environments – Benford et al., 2001; graphical interfaces and visual-symbol based interfaces; and some of the earliest HCI research assessed command languages relative to natural language learning as an everyday human competence – for example, Reisner, 1981). Such everyday competencies are deployed, however, *so as to* communicate, organise, coordinate, etc. Augmenting a door artefact would only be a sensible design choice once one understood the (local and specific) *significances* that this artefact and the associated action of "knocking" have. Sometimes what is "natural" is highly situated and thoroughly social.

For us, then, the point here is not that interaction with computation *may* be mediated through tangible mechanisms (Brave and Dahley, 1997) or through the augmentation of everyday tangible objects (such as the Media Cup – Gellersen et al., 1999) or even through natural language, speech or gesture. Rather, the key point is that the computation is in service of actions – everyday actions – which themselves have a significance. The knock on the door is an *action that signifies*. Focusing only upon the door artefact enables only a (literally) surface interpretation of what is going on and what people are doing. Augmenting artefacts needs to be in the service of both actions done with those artefacts and what is accomplished through those actions, what is "done in the doing", something that we feel resonates closely with some of the observations offered by Randall elsewhere in this book. In instance 1b the door is dispensed with completely as an artefact for coordination because that has already been done in other ways. Randall, in his chapter, cites respondents in the smart house he studied stressing that they would prefer the technology not to make things harder "than in a normal house". Similarly, here one would not want to require someone to knock on a door to announce their departure to someone who was already standing next to them ready to depart. The design goal, then, is to augment the resources, tangible or otherwise, available to the action and to what is done in the doing of that action. Put simply, we need to embed computation within life not just in cups.

10.6.3 Embedding Extra Semantics

A related approach to the one outlined above is to assume that embedding computation within an *existing* tangible artefact is guaranteed to merely "augment" that artefact in "natural" and "intuitive" ways rather than to fundamentally change (if not confuse) the semantics of exactly what that artefact is.

We have suggested then that a fundamental issue for us in things that are "invisible in use" is not the physical nature or particular perceptual qualities of these things but rather the significance which accrues to them within a particular course of action. For us, this emphasises the importance of what can be called "user semantics" and here the target is the area that is between and deliberately separate from (1) how system entities connect to each other and what they know about themselves and others; and (2) how users interact with the system through interaction mechanisms. User semantics is rather what the user makes of the computational resources (primitives, combinations, constraints etc.) and includes any accounts or representations the computational system gives of itself (Button and Dourish, 1996). That is, while we are interested in and recognise the challenges both of novel interaction paradigms and of system-level problems in ubiquitous computing, we also see a particular danger of this middle area being slipped over if issues of new user-level semantics are conflated with tangible computing interaction mechanisms – thereby failing to recognise the presence of two distinct topics here.

Not explicitly recognising this level of user semantics may make it harder to conceive and evaluate designs in which changes to the semantics of objects are being introduced. For example, one could choose to embed within a door some mechanism that displayed a personalised newspaper, or debited a credit card or changed channels on a television whenever someone knocked on it. These might or might not be desirable additions to the functionality of the door. What matters, however, is that they would change the semantics of the door, regardless of how useful or easy to learn that might be.

Furthermore, we have seen in instances 1a and 1b that *this* knock on *this* door for *these* people at *these* times is not a request to enter, not a warning before entering, not a test to detect for presence but rather an announcement of imminent departure. That is, not only is more done in the doing than just the doing but it is also the case that what is done in the doing is "just that" and not something else. Consequently while some uses of some doors by some people at some times might lead one to want to augment those doors such that, say, the doors capture details of all the people who called by while you were not there, or which displayed whether the room behind them was occupied or not, that would offer nothing to what was done with the door in instance 1a.

The nature of the augmentation is not then simply one of computation but of semantics. That extra semantics are being embedded in a tangible device is no saviour, it does not in itself render *those* semantics somehow natural. The existing semantics may be natural or at least known and understood but assigning extra semantics cannot be guaranteed to "ride on the back of" the initial semantics. Such augmentation should therefore be a matter for careful *design* reflection and indeed an artefact may have to be redesigned so as to make its new semantics understandable.

10.6.4 The Support of Everyday Routines

We have pointed to three areas in which some of the current approaches to developing "invisible and embedded" ubiquitous computing may need to be reconsidered or where the use of these approaches should not ignore the requirement for careful and prior *design* reflections.

To conclude, we want to identify one further contribution from our studies of routines and that is that routines themselves are central to the domestic arena and no doubt other domains as well. We can anticipate a growing desire for ubiquitous computing to support our everyday routine courses of action. So then, echoing concerns expressed by Cheverst et al. in their chapter, we would like to consider whether we can now look to the future and propose useful design considerations which such support systems may need to attend to if we are to practically and easily live with ubiquitous computing. What will it really mean for ubiquitous computing to fit comfortably within everyday routines and augment them without losing or disrupting the qualities that make them what they are? What design issues arise for *systems* that themselves sense and utilise knowledge of peoples' routines in order to deliver calm and *context-sensitive* support?

We have noted that in the office environment office automation systems failed to appreciate the subtlety of the status of representations of routines and the impact upon this when they became embedded within systems that constrained and determined how work flowed. Consequently we feel that it will be important for systems aimed at augmenting everyday routines to ensure that they do not transform the unremarked nature of doing routines by *marking them out* through supporting them. It could be that marking out actions within routines is the very thing that disrupts the doing of routine sequences of actions. Systems must therefore be designed such that background is not made foreground, routines are not made episodic, and the matter-of-course does not become a matter-of-comment. We should not, however, be heard as arguing that developing ubiquitous computing or context-aware computing that supports or uses an understanding of routines is therefore impossible. On the contrary, we have seen that routines *are* resources for action and knowledge of *others'* routines can also be resources for action and interaction. Routines are knowable, teachable and breachable. To some extent the same may be true for systems' comprehensions of routines. This would offer some grounds for believing that systems' may be able to usefully comprehend routines.

Design contributions may arise from understanding the details of routines – such as the point we noted in instance 2 where within many routines there are *nodal* occurrences that are implicative for the things that follow (such as the alarm clock or the knock on the door). These may be, for example, utterances that open up conversations or close them down, actions that initiate sequences or conclude them. From a ubiquitous computing point of view: Are these useful points to detect? Are they points for potential augmentation? Is an intervention that has to make these points more explicitly marked out less disruptive than another design choice?

However, we consider that the status of user accounts of routines needs careful consideration. Attention needs to be paid to the distinction between, on the one hand, routines being visibly unremarkable in their realisation and, on the other hand, accounts of routines being occasioned (with what is noted as relevant within the routine being bound up with that occasion). Put simply, users *doing* routines is different from users *describing* routines. The point then is not to deny that users can, if required, provide a description of a routine. Neither is it to suggest that this description is somehow "false" or that asking users is a "mistake". Furthermore, such descriptions may be very useful for systems to work with. Consequently this is not an argument against systems that, for example, ask users to script sequences of routine action. Relatedly, this is not an argument against systems that attempt to notice patterns of activity. As we have observed, this is exactly one of the ways in which people learn of others' routines in useful ways. However, this *is* an argument for a clear conceptual understanding of the difference between being involved in giving a description or account of a routine and being involved in doing the routine.

To take this further, it may well be that systems which intend to support the *doing* of a routine will be highly disruptive if in the course of the doing of the routine they require the user to switch to *description* activities. To do so would be to effectively pull the user away from doing their routine and to call them to account for it, to *remark* upon its elements and to thereby require an explanation of their significance.

10.7 Conclusion

We have shown how lessons that challenge and can help develop the ubiquitous computing agenda in the direction of technologies being "invisible in use" can be drawn from studying the domestic environment. In particular, recognising the subtle character of the often complex, yet unremarkable, details that surround our everyday routines places powerful requirements on any technology that might become embedded in such activities. We have provided examples that help reveal what "invisible in use" might mean but acknowledge that a great deal of research remains to be done in order to move from this to actual designs. We believe that there are deep challenges ahead in trying to provide unremarkable computing for unremarkable routines. In this chapter we have attempted to articulate some of these challenges and take a small step towards suggesting how they might be addressed.

Acknowledgements

Particular thanks are due to Jon O'Brien, Graham Button and Marge Eldridge, all of whom contributed to the ideas we have worked up here in a number of discussions. The research for this chapter was conducted in part for the MIME (Multiple Intimate Media Environments) Project, IST – 2000 26360, funded by the Future & Emerging Technologies Arm of the EC's IST Programme: FET – Disappearing Computing.

References

Benford, S, Greenhalgh, C, Rodden, T and Pycock, J (2001) "Collaborative Virtual Environments", *Communications of the ACM*, Vol. 44, No. 7.

Bowers, J (1994) "The Work to Make a Network Work: Studying CSCW in Action", *Proceedings of CSCW '94*, Chapel Hill: ACM Press.

Brave, S and Dahley, A (1997) "inTouch: A Medium for Haptic Interpersonal Communication", published in the *Extended Abstracts of CHI '97*, 22–27 March, New York: ACM Press.

Button, G and Dourish, P (1996) "Technomethodology: Paradoxes and Possibilities", *Proceedings of CHI '96, Human Factors in Computing Systems*, pp. 19–26, Vancouver, Canada, 13–18 April, New York: ACM Press.

Crabtree, A, Hemmings, T and Rodden, T (2002) "Pattern-based Support for Interactive Design in Domestic Settings", *Proceedings of the 2002 Symposium on Designing Interactive Systems*, London: ACM Press.

Edwards, K and Grinter, R (2001) "At Home with Ubiquitous Computing: Seven Challenges", in GD Abowd, B Brumitt and SAN Shafer (eds.),

Proceedings of Ubicomp 2001, pp. 256–72, Berlin, Heidelberg: Springer Verlag, LNCS 2201.

Ellis, CA and Nutl, GJ (1980) "Office Automation Systems and Computer Science", *ACM Computer Survey*, Vol. 12, No. 1, pp. 27–60.

Garfinkel, H (1967) *Studies in Ethnomethodology*, Englewood Cliffs, NJ: Prentice Hall.

Gellersen, H-W, Beigl, M and Krull, H (1999) "The MediaCup: Awareness Technology Embedded in a Everyday Object", in H-W Gellersen (ed.), *Handheld and Ubiquitous Computing, First International Symposium, HUC'99*, pp. 208–10, Karlsruhe, Germany, 27–29 September.

Henderson Jr, DA and Card, SK (1986) "Rooms: The Use of Multiple Virtual Workspaces to Reduce Space Contention in a Window-Based Graphical User Interface", *ACM Trans. Graphics*, Vol. 5, No. 3, pp. 211–43.

Hindus, D (1999) "The Importance of Homes in Technology Research", Co-Operative Buildings Lecture Notes in *Computer Science*, Vol. 1670, pp. 199–207.

Hughes, J, O'Brien, J, Rodden, T and Rouncefield, M (2000) "Patterns of Home Life: Informing Design for Domestic Environments", *Personal Technologies*, Vol. 4 , No. 1, pp. 25–38.

Lacohée, H and Anderson, B (2001) "Interacting with the Telephone", *International Journal of Human-Computer Studies*, Vol. 54, pp. 665–99.

O'Brien, J and Rodden, T (1997) "Interactive systems in Domestic Environments", in I McClelland, G Olson, G van der Veer, A Henderson, and S Coles (eds.), *Proceedings of the Conference on Designing Interactive Systems: Processes, Practices, Methods, and Techniques* (DIS '97, Amsterdam, The Netherlands, 18–20 August), pp. 275–86, New York: ACM Press.

O'Brien, J, Rodden, T, Rouncefield, M and Hughes, J (1999) "At Home with the Technology", *ACM Transactions on Computer-Human Interaction*, Vol. 6, No. 3, pp. 282–308.

Philips Research (2000) http://www.research.philips.com/generalinfo/special/ambintel/index. html

Reisner, P (1981) "Formal Grammars and Human Factors Design in an Interactive Graphics System, *IEEE Transactions on Software Engineering*, Vol. 45, pp. 59–73.

Sacks, H (1992) *Lectures on Conversation*, Cambridge, MA: Blackwell.

Suchman, L (1983) "Office Procedures as Practical Action: Models of Work and System Design", *ACM Transactions on Office Information Systems*, Vol. 1, No. 4, pp. 320–28.

Suchman, L (1987) *Plans and Situated Action: The Problem of Human-Computer Communication*, Cambridge: Cambridge University Press.

Tolmie, P, Pycock, J, Diggins, T, Maclean, A and Karsenty, A (2002) "Unremarkable Computing and the Household", Position Paper given at CHI 2002, Workshop of 'New Technologies for Families', http://www.acm.org/sigchi/chi2002/workshop-monday.html#nine, Monday 21 April 2002, Minneapolis, Minnesota.

Venkatesh, A (1996) "Computers and Other Interactive Technologies for the Home", *Communications of the ACM*, Vol. 39, No. 12, pp. 47–54.

Weiser, M (1988) "Ubiquitous Computing #1", http://www.ubiq.com/hypertext/weiser/UbiHome.html

Weiser, M (1994a) "The World Is Not a Desktop", *ACM Interactions*, January, pp. 7–8.

Weiser, M (1994b) "Creating the Invisible Interface: (invited talk)", published in *Proceedings of the ACM Symposium on User Interface Software and Technology*, 2–4 November, Marina del Rey, California.

Weiser, M and Brown, JS (1996) "Designing Calm Technology", *Power-Grid Journal*, v1.01, http://powergrid.electriciti.com/1.01

Wynn, EH (1979) "Office Conversation as an Information Medium", PhD dissertation, University of California, Berkeley.

Zisman, MD (1977) *Representation, Specification and Automation of Office Procedures*, Report from the Department of Decision Science, The Wharton School, University of Pennsylvania.

Daily Routines and Means of Communication in a Smart Home

11

Sanna Leppänen and Marika Jokinen

11.1 Introduction

In this chapter, we will discuss how the roles of family, everyday routines and communication fit in with the idea of a smart home. People are quite traditional in their way of living, at least when it comes to daily routines and chores in the household. Families have their own specific circles that are not necessarily easily pervaded by technology. There is, though, a lot of digital technology in households these days – mobile phones are actively used in communication between family members and friends, and computers are used as a means of communication, gaming and, for example, banking. There are still a lot of traditions in households, too. Television persists as the favourite media and people still go grocery shopping themselves and do not buy their carrots and milk via the Internet. Mothers still call the children in the morning instead of using an interphone. There are, one could say, huge challenges to be faced by smart home technology producers.

As an interesting example of technological content at home, we will look into how people use digital newspapers. There has been a lot of pressure to convert traditional media into electronic form. This is the case in smart house thinking overall – all that can be turned into electronic form and thus save people time, money and fuss is considered worth doing. We will go through people's ideas about this kind of a change. Do people want to read their newspapers in digital or in paper-based form? What would enhance the reading of online newspapers? How do digital newspapers fit into daily living and the home environment especially? We will compare people's ideas about traditional newspapers and digital newspapers and link it to the general discussion on a networked household.

Lastly, we will look into consumers' own ideas about the home of the future. What kinds of implications does a *smart home* have? What will daily living be like in the future? In short, we will analyse the intervie-wees' way of speaking and feeling about future technologies. People's visions of technology are often tainted by reservation and even fear. This has been the case for decades. Where does this reservation come from, even in these days of ubiquitous technology? It seems that home – as well as technology – evokes strong feelings. These feelings arise for different reasons, though. When talking about home, people see it ideally as a nest: home is safe, relaxing and comfortable. When talking about future technology, people's fears erupt. What if technology makes us cold and impersonal? Technology and home do not necessarily fit well together. That is why technology firms and smart home producers should think carefully about what kind of technological solutions they offer to consumers. How could technology be less like technology and more like home?

This chapter is based on research studies conducted in the Digital Media Institute at the Tampere University of Technology in Finland. The studies were carried out by using qualitative methods, mainly thematic interviews and some experimental methods (daily diaries, scenarios etc.). In qualitative research data is gathered by human dialogue and personal opinions are exposed into common discussion. Data is being produced in the communication between the researcher and the interviewee (see, e.g., Kiviniemi, 1999, pp. 64–65; Banister et al., 1994, p. 195; Potter and Wetherell, 1987).

In the Smart House study, which was completed in April 2001, interviewees were encouraged to imagine a future household and think about living in a highly electronic home. We were especially curious about the emotions a (future) home raises. We wanted to make daily living the starting point of our study, because that is where the smart home will work, eventually, among the daily chores and routines of people. We also wanted to explore the communication rituals in families and that is why our interviewees consisted mostly of families with children. The study of online newspapers was a study of people using both digital and paper-based newspapers. Home is not the natural environment to browse online newspapers as people use them mostly in their workplaces, during lunch hours and in order to relax from work for a while. As people see that home, in an ideal case, is a collection of people, feelings, cherished items and touchable things, digital newspaper does not seem appealing. There are, though, some advantages in digital newspapers that beat the paper-based ones. There is also a growing new generation of newsreaders who do not necessarily mind the loss of paper as they are used to using digital devices anywhere, any time.

11.2 Daily Routines Structure the Everyday Life

Smart house technologies are being developed in order to make everyday life and repetitive chores at home easier. The main objective has been to ease daily life and, especially, give inhabitants more spare time. At least this is the case in principle; it seems that technology producers still develop technology per se, and not for the real benefit of the user. Consumers have just recently become familiar with the computer and the Internet. For some, *that* has already been stressful enough. People cherish their everyday life and routines, even if they are performed in an "old-fashioned" way. There are chores that people want to do by themselves even if technology was there to help them. Routines are a part of the normal, daily life and there would be a huge gap to be filled if they were taken away. Our interviewees did not actually see the point in changing routines at home to something else. People are usually not interested in technology for its novelty but for the real benefits it has to offer. One might want some more spare time but, at the same time, many daily routines should still persist in order to keep the safe and sound life going on. Therefore, it is crucial to think about what kind of services are best left for people to do by themselves and not converted into electronic form.

Brushing teeth, making morning coffee, taking the dog out, and clothing the children, ironing, hoovering, cooking, and watching television. These are examples of daily routines that are repeated day after day, week after week and year after year – often without people even noticing them. Some of the routines are agreeable but some of them are frustrating and time-consuming. One person may like to cook but hates dusting, and the other may like to iron but dislikes hoovering. Sometimes, though, it would be nice if someone else did the daily chores to free more time for oneself, family and friends, and hobbies. This has been the overall aim of home technology, at least in images and advertisement, for decades. In the 1950s, for example, washing machines were advertised as devices that take the burden off mother's shoulders (Pantzar, 2000, p. 43). Nowadays, it seems to be a fact that technology does not increase spare time because the schedules of modern working culture and lifestyle are tight. Routines are unnoticed daily practices and they vary from household to household. There are basic routines that are the same in every household; things that must be done, no matter what: grocery shopping, cleaning, eating, drinking and sleeping. The routines may be the same but the way of performing them may vary from household to household and from individual to individual. That is why it is a big challenge to smart house producers to think of a way in which differing interests can be taken into consideration.

As Pihlajamäki (2001) points out, people do not notice doing the same routines every day, as action at home is based on routines. Only if something unusual happens, do people react to it. This fact affects people's willingness to purchase smart house technologies, too. As everything at home happens the way it has always happened, there does not seem to be much that technology can do. The only problem seems to be the hustle and bustle. Especially in bigger families, peace and quiet is often needed. Insufficient time is usually the outcome of varying daily schedules, unsystematic housekeeping and unevenly split housework.

"What would you like more time for?"

W: " . . . just for being, that I wouldn't have to think about anything for at least an hour"

M: "Me too, I would like to have more time for myself. In other words, I work too much now and that is why, in my spare time, I do not dare to be anywhere else but at home. When you think about this triangle like work, family and myself, the myself part is diminished."

Basically, what people want is more spare time. Technology has tried to offer this but it has at least partly failed. In estimating future technology that they themselves would buy, consumers are content to be rational and they highlight solutions that give clear benefits (either economic or safety-based). Smart house technologies that most people are pleased with are connected with saving energy or money. House automation is seen to be useful, including ventilation, heating and lighting. In addition to economic values, house automation could add to the home's ambiance when lights, humidity and warmth are in balance. Busy mornings are common in households: waking up is a "critical" point of a day. House automation could, for instance, ease the morning anxiety: coffee being ready, air being pleasant and warm, lights being dim enough and favourite music playing in the background. Consumers stress that homeliness and feeling good are important factors when they make choices regarding the home environment.

In the final analysis, people are rational in making technology purchases. Even if futuristic visions dwell in people's minds, in real life people do not make hasty decisions in buying home technology. As people feel strongly about their own homes, they do not want to fill them with unaesthetic technology and huge wire coils. Even technology has to fit the atmosphere. This is how a mother described her feelings about homeliness.

"What makes home a home?"

W: "Well that it looks like me, that is . . . is filled with things that I like and that look nice, a cosy, homely feeling."

"When would home feel strange to you?"

W: "If my things weren't there, if it was filled with things that are not familiar. Strange things would make me feel not at home."

"What sort of thing do you find important at home?"

W: Sofa, TV is very important, curtains, carpets."

As the example above shows, simple basic things make a home. This was the opinion of all our interviewees. The spirit of home is in people and in small, seemingly unimportant things. The smart house is seen as a practical, useful network or as a set of devices that will be worth its cost in the long run. People still think that the "heart" of the house is the relaxing combination of sofa and television. The favourite pastime at home is still watching TV. Surveys showing the penetration rates (see e.g. Nurmela, 2000) of television sets and computer terminals strengthen the fact that people do still consider TV as more important than the PC in a home environment. This is one of the reasons why smart house producers should consider centring the services to the digital television. Digital TV could be the control terminal of the smart house. Occupants could adjust the daily function of their house by a remote control (or a mobile phone). Digital television could be the terminal that combines different gadgets into one. One would get daily newspapers, TV programmes and, for example, radio plays from the same device. This kind of change would also cater to the desire for more aesthetic technology, as people say that technology should blend in better with their home decorations. There should not be too many boxes in the living room corner, appliances should look classier and there should be several designs to choose from. What would make things even better would be wirelessness. It must be pointed out that there might be a long way to go before this kind of convergence takes place in people's everyday lives. The fact is that people still consider television as suitable for just passive watching of programmes and movies. On the other hand, the majority of people are not too keen to change their present TV sets to new ones just because of some improvements that do not radically affect the actual watching. In the future, different devices are chosen according to the context rather than the content.

11.3 Family Communications

Busy weekdays and children with several hobbies prevents families from having enough spare time. That is why relationships between adults and between adults and children are taken care of, during the day, by mobile phone. It is almost necessary in the communication between a parent and a child. Timo Kopomaa's (2000) studies indicate that the mobile phone has given the parents an opportunity to "keep an ear on" their children. Children can also reach their parents conveniently if there is any trouble. The mobile phone has become a popular way to make appointments and keep in touch with family during the day. It has become very common for people to emphasise, time after time, how important it is for families to spend time together and for people to cherish face-to-face situations and social meetings.

In an ideal situation the smart home would activate its occupants. In planning home technology, producers and designers should think about moral responsibilities, too. Technology could redeem its promise and do people a favour: create more time and togetherness. This is naturally not the responsibility of technology designers only but also of administrative players. According to social theories, the late 20th century was known for its respect for individualism: life was understood as an individual project based on one's own choices. These choices cover, for example,

consumer behaviour. Making own choices means having own responsibilities. Traditional life styles and communities have nearly vanished
and the collective identity has faded (Saastamoinen 2000, pp. 161–62).
Several post-modern theorists have responded to these claims. French
sociologist Michel Maffesoli (1996), for instance, raises ideas about
how people still create communities but the way of being together has
just changed from the traditional mode. New collectivity is found in
modern gathering places, like shopping centres, coffee shops and the
Internet. People are together but still apart from each other. The fact is,
according to Maffesoli and to many others, the need for collectivity has
not disappeared. Either way, the changes in lifestyles have led to some
serious problems when people no longer have a close and tight safety
net around them. This concerns, most of all, families. A busy work culture
has led to a scheduled family life and people no longer know how to
respect communities and positive interdependency (Helsingin sanomat,
1999).

Post-modern theorists also say that collectivity is nowadays produced
by symbols rather than by interaction. Interaction has become more
superficial and lighter than before. Finnish sociologists, Tommi Hoikkala
and J-P Roos (2000, p. 26) talk about a society of weak commitments.
The Internet and the mobile phone point to this direction as, especially
younger generations, communicate a lot by these means. The more radical
thinkers like Jean Baudrillard (1994, p. 123) state that the post-modern
human being is not an independent persona but rather its counterpart
in loaning oneself to technology and at the same time exploiting oneself.
When stress and continuing scheduling of time determine how to live,
people sacrifice themselves to technology. In our study this sort of
thinking could be seen, too. Interviewees said that people's values should
be somewhere else than in effectiveness and in newest technology, especially when we are talking about home. It also seems that the society of
the mobile culture has multiplied the amount of communication but at
the same time weakened commitments. The mobile phone is a device to
make appointments but also to break appointments.

Individualism and critics put aside, smart living might, ideally, be a
gate to a new kind of collectivity. Instead of highlighting the individual
and self-dependent action, the smart house could be based on doing and
being together, on collective actions. Collectivity could include the family
members only (more quality time with children) or in an ideal case it
would spread further. Bringing suitable devices and connections to the
home might encourage the possibilities of network society and citizen
activity. Smart house appliances would enable real-time communication
with family and friends living far away, even in the form of view calls.
As we will point out later, this new collectivity will not be easy to create,
as people are still quite sceptic about technology. In the background
there is always the fear of technology taking things over and humans
becoming Baudrillardian satellites.

One very important factor in family communication is that children have more and more power to influence family decisions. This is the case in technology purchases especially, as children often know more about the latest technology than their parents. Children use entertainment technology a lot and they may raise the idea of buying a certain appliance. As children have their own rooms and their own privacy, families often need to buy appliances for their rooms, too (TV, VCR, PC, Playstation, etc.) (Chapman 1999, p. 47). To children, technology is entertainment-based but to adults it should also provide information and useful benefits. What smart home technology could concretely add to family communication is a means of organising daily living (e.g. a joint calendar) and sharing housework. It might also offer ways of doing something together (e.g. educational, interactive programs on PC or TV) and through them make living more flexible.

The fear that technology most often produces is that it might separate people from one another and reduce face-to-face meetings. Regardless of these fears this problem is usually ignored in families, at least this far. Mobile phones are used frequently but mostly for organising practical matters and to hear other family members' voices: the mobile phone facilitates family communication crucially. From the viewpoint of routines and communication people's needs define how technology is used. As it has been proven many times before, people may use technology in a totally different way than it is intended (see also Williams et al., 2000; Pantzar 1996). In our data there is an example that describes this splendidly. A young single woman describes her mornings as follows:

> I usually switch on the TV in the morning but I set it silent. I don't want to switch on the lights as the TV sheds enough light. There's no other reason than this to switch on the TV. In the morning a newspaper is enough and I don't even have time to read it properly either.

As seen above, one of the interviewees uses morning television as a lamp in the morning, as its glow is much softer than the bright light bulb on the ceiling. Users are different from one another so one solution does not suit every family and every individual. Smart house should respect home rituals and all the different means of communication. Technology may be the same but user interfaces should suit different user profiles.

11.3.1 Traditional Media in Household: Focus on Digital Media?

Home evokes feelings. Home is a human constitution, a social arena for human action. In households there are specific social norms, traditions, which frame people's actions and their everyday lives. This moral order is closely linked with the forthcoming theme: everyday routines and media use – digital newspapers in particular – in the home environment. Ideology of the smart home seen in a larger reference than just as a matter based on technology signifies from the sociological perspective as follows: the smart home concept encompasses complex socio-cultural aspects that relate strongly, for instance, with the concept of home, its meanings and social reality.

Extending the discussion from general observations made in home settings, especially in the form of everyday routines, we shall continue discussing media use in the home environment, digital newspaper use in particular, as well as meanings assigned to electronic communication, in general. Further, one of the major objectives is also to compare traditional communication and electronic communication. Digital newspapers are one separate electronic service in the Internet users' media field. Here, we give a short insight into consumers' ideas about digital newspapers by comparing attitudes between electronic newspapers and traditional, paper-based newspapers. Also, motives for using online media and context of using online media are discussed.

From the sociological viewpoint the home serves as an active interface for social reality and, in this case, it also serves as a spatial arena of media use. By clarifying briefly meanings assigned to paper mail the distinctive nature of paper will be revealed. In this respect, it is also essential to explore daily routines (including mail culture) constituted by the practices repeated inside the household in order to be able to highlight the importance of social practices in media use.

11.3.2 Home as a Spatial Arena of Media Use

The home is an extremely sensitive and private environment, which is carefully guarded by the members of the household. In reference to Durkheim's (1952, 1973) theoretical discussions on community, the home here could be seen as a social unit, a community that creates a safe social network, which is highly essential. The home reproduces specific routines and practices that uphold the whole social reality and everyday life in the household (see, for example, Peteri, 2001, pp. 5–9).

The home as a sociological concept forms a social construction based on traditional home-like elements such as warmth, safety and closeness. "Feeling familiar" is the best-known slogan for home. Another aspect relates to the idea of home being a holy place that ought to be respected; domestic peace, especially, is considered highly inviolable. This construction reproduces the organised family life, which is founded on traditional social practices. This social norm[6] of home serves as a private sphere where everyone is able to identify him or herself as the one he or she really is (see also Durkheim, 1952). According to Dovey (1985) there is

---

[6] The concept of social norm is based on something familiar and acceptable. It describes something that is socially predictable and exists in general (Fiske, 1992). It also includes generally various conventions, which reproduce and uphold the norm.

a contentual difference between "the home" and "the house". The former refers to the fundamental idea of home being filled with memories and shared experiences. The latter instead is lacking the emotional reference; a house is not capable of converting into home without people creating the social reality first (see Vilkko, 1997, 1998).

One of the prevailing distinctions of everyday routines is the cultural fact of them being based on something steady and stable. They are strongly rooted in the social reality inside the home and, therefore, it can be suggested that the home serves as an artificial interface, a social frame that reproduces these routines.

Routines that describe reading habits of traditional newspapers, for example, are strongly rooted in the spatial reality of home. Specific actions, such as drinking morning coffee and reading the newspaper, take place in the home environment in many cases only in particular places, like at the kitchen table; this enables certain social routines to be reproduced time after time. What about electronic communication and spatiality of home? The technological device itself, naturally, defines where the action, a routine, is produced. In this respect, the dependency of place (of the device) has a great effect on routines that originate from the action concerned.

11.4 Traditional Paper Versus Electronic Communication

Approaching the issue from the viewpoint of comparing pros and cons towards electronic newspapers it is necessary to investigate the factual differences between the traditional and modern media practices. The differences are strongly based on cultural distinctions that filter into people's minds and actions in the everyday media world.

Spatiality related to media use basically concerns issues dealing with the context, situations of use. Newspaper, for instance, keeps going with the receiver from place to place. On weekdays, papers are mostly read in the train or in the bus or not until after work at the kitchen table or more comfortably in the living room on the sofa. At the weekend, it is more a kind of relaxation to read papers carefully. One of the best qualities of paper seems to be strongly linked with its movable nature in its traditional sense.[7] This is certainly a great challenge for electronic mail delivery in the future. It raises, for example, the following question: how can people be recompensed for their potential "losses" in the home of the future? The desire for the distinctive smell of the ink or the rustle of the paper does not easily fade from people's minds.

There is a clear sign of people preferring paper to electronic communication (Jokinen, 2001; Koivumäki, 2001). A particular distinction of

---

[7] "Movable" in the larger reference or independence from place includes also, for instance, the Internet, which is basically supposed to offer access and availability no matter the time or place.

paper originates from its traditional nature. It is a fundamental element of communication culture; it is something people are used to and can trust. There was a strong agreement on paper being more reliable, safer and more concrete than its electronic counterpart (ibid.). Even though Finns are quite used to using computers and Internet services, especially e-mail, as a daily means of communication it is still considered a bit secondary to paper communication, which is instead considered much more official and real. The paper-based document does not lose its authenticity over time. As long as the one who is concerned keeps the piece of mail in a bookshelf it is valid and authentic. A signature or a stamp is an excellent example of authenticity or reliability of paper.

Furthermore, one of the characteristics of paper is linked with the material itself: something is real if you can touch it with your own hands and you can see it with your own eyes. Paper is a historical feature, a cultural piece of evidence.

Electronic communication has taken its toll. Not everybody is pleased with this development as the following quotation of a woman indicates (ibid.):

> I'm against all of this electronic communication, I've no mobile phone and I'll never get one if at all possible, and I'll never get one of these computer things under any circumstances, because I find all that distorts a person's sense of reality and the passing of time. Letters and the information received through mail are somehow in relation to the rhythm that is reasonably appropriate for a human being.

She sees communication technology as a kind of a threat. This demonstrates perfectly how social interaction should be seen more in a larger perspective – multidimensional – than just as a technological performance. It is a whole constructed from contentual structures of meaning, but also both form and the means of communication carry cultural meanings. The same idea can be found in the basic tenets of structuralism (semiotics): people read in messages not only contents but also the sender's intentions (Lévi-Strauss, 1967; Kunelius, 1998). In other words, it makes a difference how or in what way a friend or a relative sends a message, even if the content and the aim of the message remain the most important. The woman in the quotation above also emphasises that traditional paper communication works in the real human rhythm, whereas electronic communication alienates people from this natural rhythm and creates an illusion of something unreal and unnatural. This is connected with the idea of being constantly online, which is the most typical feature of electronic communication, and that is still considered unnatural. This has been the case for decades – there have always been people who are suspicious of technological development, its new appliances and devices (Pantzar, 1996).

11.4.1 Attitudes towards Electronic Newspapers

People do not seem to be very interested in electronic newspapers, generally speaking, especially if they have to choose between the traditional paper version and electronic version (Jokinen, 2001; Koivumäki, 2001). Similar results were found in the study conducted by Taloustutkimus (see. e.g. Nikulainen, 2001) in Finland: only 24 per cent of Finns between 15 and 74 had read an electronic newspaper or magazine during the past four weeks. Young people, however, seem to favour electronic media more than older generations do (ibid.).

The rather negative or at least suspicious attitude towards electronic newspaper use at home can be easily understood considering the context and the social aspect of the reading situation. Traditional newspapers are most often associated with relaxation and pleasant leisure activities, whereas reading electronic newspapers is instead associated with work-related matters, largely because people are accustomed to using the computer and the Internet daily at work. The social context between the home and the work environment differs in essence. At work there might be perhaps 15 minutes to check the headlines and have a short break from duties at the same time. The context in front of the screen at work is work-oriented, naturally, and therefore the actual purpose of browsing quickly through the daily rubrics of the news is mainly based on gathering information or being updated. Reading the newspaper on the sofa is mainly associated with social independence and inefficiency, instead. In this respect, the context also seems to distinguish the purpose and the objectives of the media use.

11.4.2 Advantages of Digital Newspapers

The following quotation from an interviewee describes nicely one of the best qualities of a digital newspaper: "Easy enough to use even for a yokel like me . . . a common peasant" (Consumer Research Project 1999).[8] The visual appearance resembling the printed version is seen as a factor of central importance among the interviewees, along with the ease of use and good readability. The user interface should be similar to the printed newspaper. It enables the user to glance at the front page or browse through the sections according to his or her own preferences. The familiar looking layout and the well-organised, colourful front page and general appearance give the user a feeling of being able to comprehend the medium and manage the content. The familiarity seems to be,

---

8 The study is based on interviews made among readers of a specific Finnish digital newspaper called Iltalehti Online, which has been published on the Internet since 5 October 1995. It has been the most visited online newspaper in Finland. The electronic version has its paper-based counterpart, which is also highly popular and much read in Finland.

interestingly, one of the most important factors describing the success of digital newspapers. However, it is important to keep in mind the prevailing notion that the online version of a printed medium must be conceptually different to be viable.

> I think it's really . . . really great, because it's like divided into sections. I mean, some days I'm not interested in sports, so I just skip the sport section. I just read the news and gossip. So, it depends on the day, like, you take a look at the front page and take it from there.

According to the quotation, the user felt like he was reading a newspaper, a familiar concept that is easy to grasp. In this sense, new media reflecting traditional elements is, in general, considered acceptable and desirable. A conceptual link to the traditional medium is thus essential. Another point, strongly related to familiarity, is speediness: for the user, the familiar environment (appearance) enables expeditious browsing, which is of great importance in the electronic communication world.

Usually people are able to read digital newspapers free of charge as the employer pays for the Internet connection. Browsing the Internet in the workplace is considered highly natural and it is considered to be strongly related with the work itself anyway. People are coaxed into reading at least the headlines while having a break from work. In home settings, the reading context converts from the work-oriented environment into the private and spare time-oriented environment: there is no hurry or need to take schedules or responsibilities into consideration while browsing online newspapers. Nevertheless, glancing through, for example, the newest online versions at home after work also means that the user is in charge of the action. This makes a great difference whether to prefer the work or home environment for reading online newspapers.

Motives for reading online newspapers vary but one of the general interests explaining the use is linked with the speediness of the medium. Further, people no longer need to be engrossed in the reading moment. The information on the Internet is in a simple form and it is easy to adopt. Basically, online newspapers deliver the users the newest piece of information in the short run in order to keep people updated. However, one of the disadvantages of digital newspapers could also be the basic nature of the medium – the speediness and the simplification. Is there a threat that people become linguistically lazy and impatient adopting traditional profound information, which is typical for paper-based newspapers, if they are getting used to reading only online versions of newspapers without much analytical discussion or implications?

Acting in the media world is about to change fundamentally. Still, the change proceeds slowly. Traditional social routines and adopted attitudes are rooted strongly into the everyday life and are not to be replaced so quickly. It is probable that people still prefer books and subscribe traditional newspapers even if there were online versions available. Practically, this means that no matter how felicitous or acceptable a new technological appliance such as the Internet is, it could not immediately fulfil the diverse entity of communication needs of all people or displace the old system such as traditional paper-based documents or brochures. It seems to be that two different systems are utilised side by side rather than one after the other: they serve different purposes and objectives as well as gratify different needs. Another thing related with media use, in general, is that it is based to a great extent on social habits and is not necessarily active or purposeful. People use the media because they are used to doing so, absorbing information, without giving much thought to the meanings assigned to the use (Turpeinen 1998, p. 69). This kind of ritualistic media use is of a routine nature.

Thus, the smart home idea of using new technology for gathering information or being updated in the home environment, is, generally speaking, still somehow strange for people. It raises a basic contradiction between two different contexts and social meanings assigned to these contexts: technology seems to be suitable for a work-oriented environment whereas home-orientation is based on more or less "technology-free" elements. In the end, the challenge of the smart home is closely entwined with everyday routines that cannot be bypassed but that should be taken into consideration and – especially in the implementation of the smart home converted in subtle ways into a more technological form that does not frighten people – gives them modern communication tools to achieve in the media world whatever they wish and whenever it is needed.

11.5 Emotions and Smart Home Technology

When thinking about future technology people tend to be reserved, even fearful. Future technology is usually considered as exciting – it is interesting to see where the development is going – but at the same time future technology is shadowed by many doubts. People have had reservations and fears towards technology since technology started to exist. People do not like sudden changes in their lives. The same fears that have existed for decades (even two centuries) still consume people, even though households already possess several electrical devices. It is doubtful whether one can talk about real fear or if it is just a repetition of the same rhetoric time after time. People feel the need to highlight the human pre-eminence to technology.

> The importance of human communication is in danger. The communication should not be diminished, that is the wrong direction to go. I think technology should enhance and help communication between people.

> . . . I'm afraid that people will lose their imagination, that they will no longer be creative. Children should have time to just play. Technology cannot replace the child, the parent, the cat or the dog there.

Our interviewees recounted several doubts over smart home technologies – none of which were very new to the way people react to technology. First, people do fear that technology reduces the amount of communication, as the quotations of two male interviewees above show. Nothing, according to the interviewees, should replace being together and face-to-face contacts. It would be very sad if adults taught their children to be "slaves" of technology. There is nothing that technology can offer that could replace parents' presence or cuddling a pet, for instance. It has become obvious, though, that technology has made communication between family members more lively. The mobile phone has become a natural part of daily living and communication both for adults and children, between husband and wife and between parents and children. For families, the mobile phone is much more than a phone – it can increase the sense of safety, as one is able to contact the other and be reachable at any time. Interviewees hoped for even better communication possibilities in the future. In the smart home context it could mean view phones, by which one can contact friends and family who live far away. There is a fear, though, that the implications of communication technology at home turn out to be the opposite: Patricia Wallace (1999), for instance, claims that technology separates people from one another inside a household.

Secondly, technology is seen as passivating. This is particularly true when talking about the home. If technology is developed to do all the chores on behalf of people, is there any worth to humans any more? The protestant ethics still seem to affect people's opinions. Chores should be done and laziness avoided. It is still true that daily routines create social stability and safety. People are not apt to giving them up totally. Chores are a part of children's education, too; participation teaches them responsibility. Electrical appliances could do all the unpleasant chores at home but there is still, however, some housework that people enjoy doing. Time that you save from doing housework might very well be spent relaxing, in self-development and family time together but interviewees often say that time could already be more wisely scheduled if it really were that necessary.

> If I should live in a house like this I would, at some point, go out in the garden and leave all "communication buttons" on the table. One should not be reachable at all times, I think it's awfully stressing.

> . . . this seems very unsafe. If, for instance, a microwave oven was connected to the net, there is a possibility that an outsider hacks into your system and starts the oven without you knowing about it. I don't think that all the devices should be connected to each other.

Thirdly, there is the fear of technology mastering man. This point might seem a bit ridiculous and old-fashioned but as the quotations above make clear, people still state the same worries. In many of our interviews, this was an ever-present worry. Not only in the sense of science fiction movies (*Computers will take over the world*) but in talking about very concrete situations. If a home were smart, it would be wired. Wired means that all the devices are connected to each other. Here rose the scepticism: what if the home starts living a life of its own? What if, in a problematic situation, there is nothing a layperson can do to solve the problems? Does the home stop working all together? People do not trust computers or networks. It is worthwhile being sceptical about information security: would it be possible for someone to hack the home network from outside the house? Home is a place where one wants privacy and relaxation, not the idea that "big brother is watching". The interviewees who had most experience in using information technology – quite often they were also working in the ICT sector – reacted to problems of information security. Even if people are interested in different devices and smart house solutions, they are quite sceptical about their functionality. In the "horror scenarios" that the interviewees envisioned, one could perceive a hint of amusement. Interviewees wanted to imagine a home where everything is wired and where devices are everywhere, and then they laughed and were horrified what life would then be like. People could just turn into total couch potatoes as technology did everything else around them from hoovering to washing dishes and doing the laundry.

In addition, people think about technology as something inevitable. Technology seems to be, in speech, an untouchable entity, beyond man's control. The common attitude is: "The amount of technology will increase in the near future, no matter what you do. You will just get used to it, in the long run". People have the tendency to think that technology is the answer to almost every question. Solutions have been sought to health questions, interaction, safety etc. As Pantzar mentions in his book about the home of the future (2000, p. 192), digital devices are nowadays even expected to work as "life management tools". According to technology visions, people would transfer working (and thinking) to the machines. These visions have long existed and yet one cannot say that they have become true, at least not in the most negative sense.

11.6 Conclusions

Living in a networked home with lots of new technology contains many contradictions. Smart homes should contain both the option of privacy and the option of linking locally and globally. Home should be a private nest but also a public arena. There is also the danger of losing one's privacy, when occupants could be observed from outside the house. This means for example services, when a user leaves a mark every time he/she logs in (either on a PC or digital television). In the worst scenario, from the inhabitants' point of view, a smart house would be a modern Panopticon, introduced by Foucault, because the need – and threat – on control would be emphasised both inside and outside the house (Allon, 2001). On the one hand it would be magnificent if the house managed to take care of home automation, safety and expenses, but on the other hand people are too scared to give the control to technology alone. It is important that some sort of control remains in inhabitants' hands. Using information technology affects people's concentration level as it is, because it is quite hard to find relevant information from all the information available. In addition to this information overload there is the constant need to be accessible and "online". The stress symptoms may not show yet but they are due in the long run (CNN.com, 2000; Delio, 2000.) One can find contradictions also in the collectivity that is maintained by technology. In practice it can mean that people part from each other and from the world. Technology makes it easy to avoid eye contact. This has been discredited to some point by a UCLA study that claims that the Internet does bring friends and family together, that the Internet is becoming the new family heart (Christian Science Monitor, 2000). In Finland, studies show, though, that people consider computers and the Internet as highly personal communication devices. Only in exceptional situations, they are used together with others.[9] Individual differences vary in this, too.

As the future is insecure, people are reserved about future technologies. At the same time, though, people are interested and enthusiastic about technology. This is how it has been for decades. An interesting point is that technology raises very extreme emotions: fear and exhaustion, enthusiasm and activity. The so-called early adopters will probably be more willing to have more devices at home as they are not that afraid of networked technology. Most people seem to be, though, quite traditional in thinking about home environment, and it seems to be hard to combine "technology" with "home", even if people already have a lot of technology at home. Routines and traditions affect the fact that people attach deep feelings to doing things as they have always been done. This can be seen in the comparison between traditional newspapers and digital newspapers as well as in the comparison between paper-based and electronic communication in general. People prefer paper-based versions to digital versions. There are many reasons for this: touchability, familiarity, mobility, smell, etc. and probably the most decisive one is familiarity, emotion-based action. There is an own place for digital newspapers, too. Digital means of communication have not come to replace traditional means. They are an addition. People read digital newspapers when they cannot reach paper-based versions. Digital newspapers serve as a pastime in a different way to their paper-based counterparts. They are usually read at work, as a breathing space in between work tasks. Digital newspapers are hastily browsed and therefore it is not important that they include all the same information as the paper-based versions. When

---

[9] This point has risen in several case studies done by the Consumer Research Project.

people want to really concentrate on their reading, they choose traditional newspapers and traditional magazines. All this is naturally due to digital newspapers and magazines being relatively new means of information. Paper-based newspapers and magazines already have rituals around them. It is therefore interesting to see how things will develop in the long run. Will digital newspapers become steady parts of information retrieval?

There is no surprise in what people want from a smart home: saving money and energy and having more spare time. People are not ready to change their daily practices into something else. People like spending time doing chores and having concrete things in their hands. Technology can be of help but a satisfactory compromise must be found. It is obvious that not all people will or can benefit from smart living. It will long be a privilege for those who have enough money, who know how to use information technology and who have fairly new houses or apartments. All in all, it became clear in our study that technology cannot create homeliness. Inhabitants themselves make a home and little everyday practices make the known life go on. People also want to make decisions concerning their use (or non-use) of technology themselves. That is why there should always be the possibility of managing everyday life both with and without technology. Technology should adjust to people's needs and wants as any consumer product, without forgetting that not all users are technology freaks and early adopters. Home is, ideally, a secure, warm and beautiful place with loved ones that should not be spoiled with demands of control and activity. A smart home should not be smarter than its inhabitants.

References

Allon, F (2001) "An Ontology of Everyday Control: Space and Time in the 'Smart House'", paper presented at the "Spacing and Timing: Rethinking Globalization and Standardization" conference, Palermo, Italy, 1–3 November. http://www.emp.uc3m.es/~quattron/conference/papers/Allon.pdf

Banister, P, Burman, E, Parker, I, Taylor, M, and Tindall, C (1994) *Qualitative Methods in Psychology: A Research Guide*, Buckingham: Open University Press.

Baudrillard, J (1994) *The Illusion of the End,* Cambridge: Polity Press.

Chapman, T (1999) "Stage Sets for Ideal Lives: Images of Home in Contemporary Show Homes", in T Chapman and J Hockey (eds.), *Ideal Homes? Social Change and Domestic Life*, London: Routledge.

Christian Science Monitor (2000). "Internet Becomes the New Family Heart", by P Van Slambrouck, 26 October, http://www.csmonitor.com/durable/2000/10/26/fp2s1-csm.shtml

CNN.com (2000) "UCLA Study: Internet Doesn't Isolate Most People", 29 October, http://www.cnn.com/2000/TECH/computing/10/25/internet. study.ap/index.html

Delio, M (2000) "Cubicle Blues Blamed on IT", *Wired*, 11 October, http:// www.wired.com/news/technology/0,1282,39406,00.html (14 August 2001).

Dovey, K (1985) "Home and Homeless", in I Altman and CM Werner (eds.), *Home Environments: Human Behaviour and Environment,* New York: Plenum Press.

Durkheim, É (1952) *Suicide: A Study in Sociology,* London: Routledge.

Durkheim, É (1973) *On Morality and Society: Selected Writings* (ed. RN Bellah), Chicago: University of Chigago Press.

Fiske, J (1982) *Introduction to Communication Studies,* London: Methuen.

Giddens, A (1994) "Living in a Post-traditional Society", in U Beck, A Giddens and S Lash (eds.), *Reflexive Modernisation Politics, Tradition and Aesthetics in the Modern Social Order,* Cambridge: Polity Press.

Helsingin sanomat (1999) Professori Riitta Jallinoja: työelämän kilpailuhenki vallannut kaikki elämänalueet. Kotimaa, 20 November 1999.

Hoikkala, T and Roos, JP (2000) "Onko 2000-luku elämänpolitiikan vuosi-tuhat?", in T Hoikkala and JP Roos (eds.), *2000-luvun elämä. Sosiologisia teorioita vuosituhannen vaihteesta,* Tampere: Gaudeamus.

Jokinen, M (2001) Postin elinkaari – historiasta nykypäivään, kynnys-matolta kansioon. In: Digitalisoituvan viestinnän monet kasvot. Consumer Research Project's Research (unpublished).

Kiviniemi, K (1999) "Toimintatutkimus yhteisöllisenä projektina", in HLT Heikkinen, R Huttunen and P Moilanen (eds.), *Siinä tutkija missä tekijä, toimintatutkimuksen perusteita ja näköaloja,* Jyväskylä: PS-viestintä Oy.

Koivumäki, M (2001) "Älykoti ja perheen viestintä", in *Älykäs koti,* the Consumer Research Project's Research (unpublished).

Kopomaa, T (2000) *The City in Your Pocket: Birth of the Mobile Infor-mation Society,* Helsinki: Gaudeamus.

Kunelius, R (1998) *Viestinnän vallassa,* Juva: WSOY.

Lehto, M Talonpoika, R and Huovila, P (1993) *Älykäs asunto – tietoyhteiskunnan koti.* VTT tiedotteita 1457, Espoo.

Lévi Strauss, C (1967) *Structural Anthropology,* Garden City, NY: Anchor.

Maffesoli, M (1996) *Time of the Tribes: The Decline of Individualism in Mass Society,* London: Sage.

Nikulainen, K (2001) Joka neljäs lukee lehtiä webissä. http://www.digi-today.fi/digi98fi.nsf/pub/md20010612104227_kni_57752803

Nurmela, J (2000) *Mobile Phones and Computers as Parts of Everyday Life in Finland,* Helsinki Statistics, Finland.

Online Media in Everyday (1999) Consumer Research Project's Research (unpublished).

Pantzar, M (1996) *Kuinka teknologia kesytetään.* Kulutuksen tieteestä kulutuksen taiteeseen. Helsinki: Tammi.

Pantzar, M (2000) *Tulevaisuuden koti.* Arjen tarpeita etsimässä. Helsinki: Otava.

Peteri, V (2001) "Mistä syntyy kotoisuus?", in *Älykäs koti,* the Consumer Research Project's Research (unpublished).

Pihlajamäki, T (2001) "Arkirutiinit, tottumukset ja tavat", in *Älykäs koti,* the Consumer Research Project's Research (unpublished).

Potter, J and Wetherell, M (1987) *Discourse and Social Psychology,* London: Sage.

Saastamoinen, M (2000) "Elämäntapayhteisöt ja yhteisöllistämisen teknologiat – identiteetti, ekspressiivisyys ja hallinnointi". in P Kuusela and M Saastamoinen (eds.), *Ruumis, minä ja yhteisö,* Kuopion yliopiston selvityksiä.

Turpeinen, P (1998) "Tarpeet ja motiivit", in *Uusmedia kuluttajan silmin.* Digitaalisen median raportti 2/98, Helsinki: Tekes.

Vilkko, A (1997) "Ikääntyminen, muistot ja koti", in A Karisto (ed.), *Vanhuus kaupungissa. Miina Sillanpään syntymän 130-vuotis juhlakirja,* Juva: WSOY.

Vilkko, A (1998) "Kodiksi kutsuttu paikka. Tapausanalyysi naisen ja miehen omaelämäkerroista", in M Hyvärinen, E Peltonen and A Vilkko (eds.), *Liikkuvat erot. Sukupuoli elämäkertatutkimuksessa.* Tampere: Vastapaino.

Vilkko, A (2000) "Riittävästi koti", *Janus,* 3, pp. 213–30.

Wallace, P (1999) *The Psychology of the Internet,* Cambridge: Cambridge University Press.

Williams, R, Slack, R and Stewart, J (2000) *Social Learning in Multimedia.* Final Report of EC Targeted Socio-Economic Research. http://www.rcss.ed.ac.uk/research/slim.html

Living Inside a Smart Home: A Case Study

12

Dave Randall

12.1 CSCW and the Domestic Environment

Ethnomethodologists have observed that a striking feature of the sociology of work is its lack of interest in anything that looks like work itself. This absence goes some way toward explaining why it is that it is ethnomethodology, of all the sociological perspectives, that has forged a link with "design", above all in the research arena of computer-supported cooperative work (CSCW). This is precisely because ethnomethodology has taken the phenomenon of "work" itself to be a topic of its enquiries.

There is a small irony, then, in the fact that recent research begins to move enquiry away from traditional workplace studies and into public and private spaces. This is no surprise, for it associates with the spread of mobile telephony and wireless information devices, tele- and home-working, assumptions about information use for private purposes and a general interest in what the affordances of new technology might be in settings such as the home. The technologies in question also support interactional behaviours that cannot be termed "work" at all, at least in the sense meant by, for example, Schmidt (1991), although arguably more tractable to the ethnomethodological conception (see Hughes et al., 1992).

The concern in this chapter, then, is to contribute to the extension of CSCW interests into domains which are not to do with workplaces by examining research into new technology and domestic environments conducted in a so-called "Smart House".[10] In fact, research into technology and domestic environments can be traced back at least to the broadly Taylorist interests in the kitchen associated with Lilian Gilbreth (1927), to feminists interested in the relationship between technology and domestic work (see, for instance, Berg, 1994; Bose et al., 1984; Cowan, 1983; Vanek, 1978; Wajcman, 1991; and Cockburn 1997), and more recently to work by Mateas et al., 1996; Barlow and Gann, 1998; Hindus

---

[10] Although work may well go on in the home, of course, namely the phenomenon of "tele-working".

(1999), O'Brien et al. (1996) and so on (see Aldridge, Chapter 2 of this book, for a complete overview).

Having said that, there have been relatively few opportunities to examine family life as it occurs in an already existing "smart house". The specific relevance of CSCW to the arguments offered below is that, regardless of its historical interests, it offers the notion of *interactional affordances* (Martin and Bowers, 1999; Harper et al., 2000) as a major feature of the success or otherwise of new technologies. In this way, the design of new technology was held to be dependent not only on issues of *usability* (associated with laboratory-based measures of human-computer interaction), but also on *usefulness* (associated with the way new technology might be woven into people's real life and real-time experiences at work). It should be obvious that such fundamental issues translate straightforwardly into the domestic realm. The usefulness, for example, of new broadband technologies will equally depend on the real world, real-time behaviours and interactions of people.

Hindus (1999) also calls for more research into homes and technology on the grounds that they are economically too important to ignore, and have the potential to improve everyday life for millions of users. Nevertheless, Hindus argues that research specific to workplace settings cannot easily be generalised to the home context. As she points out, homes are not typically designed to accommodate technology, they are (typically) not networked, nor do they have the benefit of professional planning, installation and maintenance of technology and infrastructure. Equally, "consumers are not knowledge workers"; motivations, concerns, resources and decisions are different in the home. Thus, where workplace purchasing decisions are determined by concern with productivity, householders may well be interested in matters such as aesthetics, fashion and self-image. Further, "families are not organisations"; they are not structured in the way that corporate organisations are structured, and decision-making and value-setting are quite different. These two points go some way toward explaining why, as Venkatesh (1996) has suggested, "More segments have opened up in the 1990s, signifying greater impact and diffusion of computer technology in the daily life of the household" (p. 51). If so, the orthodox concerns of CSCW with work and interaction might be developed in such a way that we begin to understand the ways in which domestic and work environments are both similar and different.

12.1.1 Some Initial Problems

The first and most obvious difficulty we have in the examination of the relationship between family life and new technology is that evidence about the routines of family life is surprisingly hard to come by. The evidence we do have comes from one of two sources. It is either "broad-brush" and dealing with economic relationships (cf. Hamill, Chapter 4), or it is more or less orthodoxically sociological and dealing with the relationship between new technology and the structures and processes of the family. The history of the social sciences is replete with explanations of structural changes in family life, including arguments about changing gender roles, the move from extended to nuclear families, and subsequently towards "marginal" conceptions of the family, such as that of the "single parent" family.

Sociological arguments about the typical shape of the family are interesting only insofar as they relate these shifts to changes in the industrial and commercial fabric of the nation. The general move toward the small, nuclear family was conventionally held to be associated with the rise of capitalism, and particularly factory production.[11] If so, then, one could feasibly see the move toward post-industrial, post-Fordist production as also carrying implications for the family. This may prove especially true of broadband communications technologies, as they offer more opportunities for teleworking, implicate the extension of the working day into periods of time traditionally associated with family life, and generally offer substantially enhanced opportunities for connectivity with wider networks. In any event, attention to family structure can at least point to some themes that may turn out to have a bearing on the use of new technology.

Whether or not we have seen a move toward nuclear and even smaller family units, it is clear from the results of the study reported below and from elsewhere that in many respects the extended family is alive and well. That is, regular contact with a wider and dispersed set of family members should be regarded as a typical feature of modern family life, arguably more so with the advent of widespread communications technology such as e-mail. In a sense, of course, 'twas ever thus given that the letter and telephone have existed for a long time now. Even so, mobile telephony and text messaging, digital images and video, netmeetings and so on, all afford regular contact with others, not only on an individual basis, but also collectively. There is some evidence from our study that there is already take-up of these possibilities. A significant feature of this, however, is that such kinship connection does take place only at a distance. A common way of expressing family values is through the ordinary rituals of life, including births, marriages, Christmas and other festive occasions.

Second, there are problems with conceptualising the "smart house" itself – it seems that there is relatively little agreement about what a "smart house" might be, and what its relevant technologies would look like. A starting point might be the distinction drawn by Gann et al. (1999)

---

[11] This argument has been controversial in sociology. It has been pointed out that the rise of the factory system actually led to an increase in the extended family work as the need for child-care became more prevalent (Laslett, 1972).

between homes that simply contain smart appliances and those that allow interactive computing in and beyond the home. Maintaining Gann et al.'s focus on the functionality available to the user, we might identify five types of smart home:

1. *Contains intelligent objects.* The home contains single, stand-alone appliances and objects which function in an intelligent manner.

2. *Contains intelligent, communicating objects.* The home contains appliances and objects which function intelligently in their own right and which exchange information among themselves to increase functionality.

3. *Connected home.* The home has internal and external networks, allowing interactive control of systems, and access to services and information, from within and beyond the home.

4. *Learning home.* Patterns of use are recorded, and the accumulated data are used to anticipate users' needs. See, for example, the Adaptive House (Mozer, 1998), which learns heating and lighting usage patterns.

5. *Alert home.* The activities of people and objects within the home are constantly registered, and this information is used to anticipate users' needs. See, for example, the Aware Home (Kidd et al., 1999).[12]

The "smart house" discussed below is in most respects an example of (3), the connected home. As we shall see, the research detailed below supports the idea of moving towards (4) at least.

12.2 The Study

The study took place at the request of a large provider of mobile telephony services. The project entailed the building of a functioning "smart house" in which new domestic technologies could be evaluated. Because the house in question is a research environment, it has no permanent residents.[13] Attempts to evaluate the technology-in-use in the house, then, had to be done through the evaluation of short-term family residence.

The house is characterised primarily by three elements. First, there is a set of elaborate control mechanisms with which technology in the home can be managed by family members. For a mobile telephony company, the possibility of effective use of technology in a location-free way has huge potential, and thus the use of control devices, including mobile phones, was a major feature of the evaluation we conducted. Indeed, seven distinct methods of control were, in principle, available to visitors,

---

[12] I am indebted to Frances Aldrich for this typology.
[13] Not least because it periodically undergoes radical change as new technologies and systems are installed, and old ones removed.

including wall-mounted control panels, Compaq TP/IP devices, and mobile phones. A second feature of the home is the use-technology, all of which was commercially available and included sophisticated entertainment media, kitchen equipment, baby monitoring, computer networks, security systems, and so on. A third feature is the provision of various facilities which could be used in conjunction with the available technology, such as a health monitoring service and Internet shopping.[14]

The methodology entailed an ethnographic orientation, obeying the injunction that, just as with working life, the point was to try to understand domestic life from the point of view of those living it. To this end, video recording was done continuously in all "family rooms" in the house. Family members were "shadowed" through the house on an occasional basis, and were interviewed at the start and finish of each period of residence. At the outset it should be emphasised that only three families have had an opportunity to be resident in the house, and only for limited periods of time (the longest being two weeks). The families are similar in many respects, in that all have "professional" fathers and part-time working mothers, and in all of them at least one parent could be regarded as highly computer literate. All three families had at least some familiarity with the Internet, with chat rooms, and with digital and video imagery. All were familiar with mobile phone technology. Individual members of the families had expertise with MP3 and Midi, netmeetings, Search engines, digital video and photography. Two of the three fathers had sophisticated understanding of the use of electronic resources for music production. Each family was structurally broadly similar, in each case having three young children. The oldest child in any of the families was 12. No teenagers, elderly people, or extended kin formed part of the study. In two of the families the parents were in their 40s, and in one, their 30s.

12.2.1 The Practicalities of Family Life

As mentioned, sociological work on the family makes little reference to the practicalities of family life. I begin, therefore, by suggesting what some of the ordinary exigencies of family life might be.[15] First, *control* seems central to family life in a number of senses, including both control over the house itself and control as manifested in relations between family members. Second, it includes elements of what can be called *social*

---

[14] Internet shopping is, of course, a major topic in its own right. I do not report on it here other than to say that families were allowed to shop for groceries at no charge to themselves (up to a certain point) as long as they used the Internet to do so. Their general negativity, therefore, was striking.

[15] I should stress that these are analytic glosses and that much of what is reported on could be thought of in terms of any and all of these categories.

connectivity, which simply means the normal desire of family members to be in touch with each other and with a wider network. We can also distinguish between local and distance connectivity – the former being the way in which family members group together or not in order to complete various activities, the latter referring to the way family members are also outward-looking, using technologies to relate to wider networks of friends and kin. Third, *location* is of evident importance to family members. By this is meant the extent to which family life is conducted in quite specific locations. This is not the first study, for instance, to observe the degree to which the kitchen can be a locus for family-oriented activity. These themes are examined below with reference to the control systems, use technologies and other facilities mentioned above, and then some general conclusions are drawn concerning the issues likely to prove germane in the future.

12.2.2 Control

All visitors to the house were reminded that its facilities were not to be seen as final versions, but as interim solutions, the purposes of which were to elicit reaction. It was interesting, therefore, to see the dimensions that governed responses. At the outset, one should stress that the ability to control a range of functions remotely was very positively received. The ability to control some functions was clearly a great pleasure in certain circumstances: "I always read in bed, and it's nice to just reach over and switch the lights off. Same in the morning – I have trouble getting up – it's lovely to be able to open the curtains from bed." This sentiment was repeated by more than one person, and applied equally well to not having to get up from the sofa.

Equally interesting, however, were negative sentiments. These centred on problems associated with overhead, robustness and reliability, and (ironically) lack of control. Overhead here is not to be equated with cognitive load. People often reported no difficulty using control systems, but nevertheless expressed intense irritation. "Overhead" here refers to whether the ecology of the setting is such that people can complete tasks in a simple, elegant way or not. It seems that what we observe in work settings (see Harper et al., 2000) is paralleled by domestic life, down to the fact that unnecessary seconds seem to matter to people in domestic settings as much as at work. A simple example of this issue was the lighting in the house. The existence of an overhead in doing simple things like switching lights on and off was a constant irritant. Comments such as the following were typical: "Things must be simpler to do than in a normal house . . . I don't want to work through a menu just to turn off the lights. Again, I hope this will be improved with voice control", and, "It should never take longer than it did before. Keep it simple". Similarly, "we need manual over-rides. We do not want to fiddle with remote control for the washing machine when we're standing in front of washing machine . . . " and, "The controls just aren't sophisticated enough to run the washing machine, and do you really want to spend five minutes trying to get it to do what you want?"

Similarly, robustness and reliability turned out to be significant. Examples abound in the study of family members experiencing, for them, strange and bizarre behaviours by the control devices. As one put it: "The plasma screen was completely unreliable – the only way I could get it to switch on was by re-booting the control device. The DVD facility was the same – the only language we could get the DVD to play in was Danish! If I went through the wall panel and selected DVD, then the device would work with the Plasma screen. That's really weird . . . " For one family, system unreliability culminated in a minor disaster: "The cupboard doors between the bathroom and the master bedroom were stuck open. We couldn't get into the bathroom at all. And the control device was saying the doors were closed!"

This issue of reliability was nowhere more evident than with the locking and security features. Again, it should be stressed that the general principle of security systems of this kind was very warmly received, as in, "I really liked it. I felt very secure. I think its very good to be able to check up that you've locked all the doors and windows from afar . . . ", but unreliability was a critical factor: "I felt that there was a real risk that people would get locked out. In fact, while I was there the kids got locked out in the garden because there are no door handles on the outside of the patio doors in the kitchen." Similarly, "We went out once and I locked up, and I decided I just needed a wee before . . . and I went back into the house but I couldn't get back into the bathroom . . . "

Perhaps most interesting, however, was the paradoxical sense in which elaborate control mechanisms could generate a sense of lack of control. By this I mean that control systems were resented if they did not allow users to engage in and complete the activities they wished to undertake, and where designers had simply presumed they could predict what users wished to do. Thus, some users (though by no means all) expressed negative sentiments about the bathroom: "The bath, though, it didn't fill up off the control panel and it's a daft idea anyway. Actually, the bath in the en-suite bathroom doesn't empty properly either. I can't imagine why anyone would want to run a bath remotely." When asked whether there might be specific benefits for, for instance, the disabled, one father agreed, but said, "I'm still not sold on it, even if it was tailored to my specific desires. I like bubble bath and you can't put that in afterwards . . . there's always contingencies, and you can't do it. The top-up button doesn't work well enough. I wasn't getting the control I wanted." The same adult went on: "Simple tasks just look a lot more complicated. I left a room, switched a light off, went out and then remembered that I'd left something in there and was fumbling around in the dark", and, "There's not an ordinary tap in the house and it drives you mad. You can't control the water

volume and it's inconsistent. I really disliked the *lack of control*." Paradoxically, it seems, the elaboration of control can result in a sense of lack
of control.[16] A potentially important element of this derives from the fact
that family members do not naturally check to see if others might also
be interacting with the control systems: "We've already discovered we
find ourselves all trying to control the same thing at the same time. They
(the control systems) don't tell you that someone else is trying to do the
same thing. Overall, it's got to be quick and simple." This lack of feedback may be in part why all the families reported odd "mysteries", where
things did not happen in quite the way anticipated.

12.3 Social Connectivity

12.3.1 Local Connectivity

Previous research into domestic life has tended to emphasise the importance of location. There is no need to demur here, for there can be no
question that family life is currently location-oriented. That is, the use
of certain technologies is normally associated with specific activities in
specific places. The video data confirm other research in making it clear,
for instance, how important a place like the kitchen can be for family
life.[17] Nevertheless, video observation also provides us with a rare insight
into the *rhythms* of family life. One feature of this, seldom remarked
upon, is the way in which families are rarely "collective entities" for long
periods of time. They come together on certain occasions, for example,
when eating, but soon separate to engage in more individual pursuits.
Nevertheless, they remain families. That is, even when engaged in isolated
pursuits, family members regularly "check out" the activities of others.
The video data show that husbands and wives, when both home, though
perhaps engaged in utterly different activities in different parts of the
house, will move back and forth for short periods.

This is even more marked with the behaviour of children. Children
are often a locus for the most pertinent of the "privacy" versus "connectivity" issues that are central to family life. Younger children, as we all
know, will "pester" older children to play. They sometimes appear to be

---

[16] This sense of lack of control was evident in ergonomic matters as well, reflecting the need
for attention to particular categories of user: "Little kids can't reach the control panels, and
they need lights to do things like sit on the potty. The cleaner, Mary, had to borrow specs to
read what the control panels said." (general laughter from a rather middle-aged group of people).
[17] One feature of the house as initially designed was the way in which assumptions were made
about appropriate technology in the kitchen. Thus, where technologically advanced fridges,
washing machines and dishwashers were all present, the absence of other technologies was
keenly felt: "It's typical, really . . . there's no decent TV here . . . this kitchen was designed by
young designers for whom looks were more important than function and who were not familiar
with family life with young children."

joined by invisible strings to their mothers (video data showed how frequently children appear in the kitchen for brief periods when mother is working there. Occasionally, they remain near and engage in play in the kitchen).[18] This local social connectivity has a number of repercussions. A significant finding of the video data was the constancy of "monitoring behaviour" in the family. Parents, it seems, habitually check up on the whereabouts of their children and each other. Moreover, it is common for the children to engage in the same behaviour toward each other and toward their parents. Part of this will be normal parental anxiety, as expressed by one mother when she said, "I need to keep an eye on Peter, who has a bit of a tendency to run off". This kind of behaviour goes a long way toward explaining the popularity of "surveillance" technologies like the baby monitoring equipment in the house.

Questions concerning what technological affordances are appropriate to what locations, or conversely whether developments in computing might make domestic technology location-free, will be answered in part through understanding the nature of these rhythms. One can usefully describe this issue as being on a continuum from "personalisation" to "integration". It is a truism that not everyone has the same priorities, and typically with the use of, for instance, PCs we are prone to "individualising" or "personalising" the technology. Indeed, there was evidence of the importance of such things for young people through one 12-year-old's desire to spend time online in order to download various pieces of software that he could use to personalise his mobile phone. These issues are not only important to the young; we see similar demands with kitchen technology. At the same time, personalisation of technology is a risk, in that the more personal the interface, the less usable it is by others. This is a particular risk in the context of family life, and was evidenced in data around the use of, for instance, the CD system. The other pole, then, is that of *integration*, whereby all functionalities can be used by all family members everywhere.

Integrated technologies proved very popular in certain respects. Most adults found uses for controlling devices at one location in the house from another: "Switching music on from anywhere in the house is great" This did not, however, apply to devices where physical presence would at some point be required: "I have to go to the dishwasher to load it, so I don't need a remote if I'm in the house. I could see a use for a coffee machine, so coffee was ready when you get home. Can't see why I would want to do the washing remotely". Surprisingly, there was general scepticism about remote access, which had to do mainly with fitness for purpose: "Why would I want to run the washing machine from outside? I think you'd have to be very fussy to care . . . I suppose there are some people who don't like leaving clothes in the washing machine because

---

[18] I should perhaps note here that I mean no sexism. In all the families studied, mothers were homemakers, or worked part-time, whereas fathers worked full-time.

they get musty . . . " Similar observations were made with regard to cookery and related functions: "The longest thing we buy to cook is stuff like frozen pies. I suppose I might want to turn the oven on . . . I might . . . but it'd be pointless really, because the pies would be defrosted anyway, wouldn't they?"

In any event, the argument is for a more nuanced view of location, one which takes account of whether activities are individual or cooperative, and the occasions on which this may be the case. Distinguishing between the two is no trivial matter. Thus, adult information-seeking behaviour in the "smart house" tended to be something that parents did alone and at night. Two out of three families broadly followed this pattern, especially where the "surfing" activity in question was hobby-related. The main exception to this occurred with highly specific and short-term information seeking. One family, for instance, showed us how they had decided to buy Chinese food one evening, but being unfamiliar with the area had used a Mapping service to identify exactly where it was.

Entertainment, and especially television use, seemed to have a more complex patterning. On the one hand, older children reported that they watched TV on their computers to "get away" from their siblings. Equally, the study was used as a place to get away from the kids by most of the adults. One mother spoke of watching on her own in the bedroom: "I pretty much always watch TV just before I go to sleep . . . " This suggests that some technology can be personalised by location, particularly where functionality needs to be allocated for people working from home. (Video data shows that the one father who did any work at home did so in the study).

It is also entirely in keeping with what we know about the spread of second and third TVs and videos through the home. Interestingly, when children watched TV on their PCs they invariably did so on their own. Having said this, I do not want to give the impression that TV watching has become an entirely individual phenomenon. Films, in particular, are sometimes an occasion for all the family to sit together and watch, and it is normal to have one TV set which is co-opted for family use. Thus: "We don't watch a lot of TV together, but we have a widescreen TV in the back room, with Sound-Surround. Sometimes, we'll all watch a film – we sit and watch all the way through it." In another family these facilities also proved very popular with the children. Hence, "The kids liked the fact that you could watch TV and use the computer on the same screen – they could switch from one to the other. In fact, though, the kids watched a lot of TV in the adult bedroom on the Home Entertainment System . . . because of the screen quality. All three of them would be in there playing with the bed settings and watching films . . . We did find we were rather less likely to watch as a family". One father had a rather different view: "I can't see us using these Interactive TV facilities a lot. We have one main TV at home, and there's already too much dispute between the kids." The lesson to be drawn from this is that rather than location itself, the individual or collective purposes of family members is the critical issue.

Research (Hamill, Chapter 4, this book) has also suggested that online education is becoming an increasingly important reason for logging on. If so, one can argue that we need to know a great deal more about what the educational activity in question is. If variation is to be found with something as commonplace as TV watching, it is likely that it will be found elsewhere as well. One area where this proved to be of particular interest in the study was that of information use and educational activity on the part of young children. Parental involvement in educational activities with small children was striking, largely because the children themselves demanded it.[19] That is, educational activity for young children is typically cooperative. In practice, it often requires parental or sibling input: "Computerised tasks seem to take a long time, and we tended to find we were occupied with all sorts of things. How long did it take you and Sam to do those invitations?" – "it took over an hour. Steve had to start the dinner. That was because it was the first time we'd used it, though. We did enjoy it, Sam especially." This is in keeping with other research in the educational arena which also stresses the importance of timely adult intervention with interactive media (Hemmings et al., 2001).

This was acknowledged by the children as well: "Lee helped me to find wallpapers and the Buffy screen – its hard to find ideas on your own. My mum had to show me Photofun, and you could move around in it . . . We printed things off the [live]board. It was good, I like the board."

Thus, understanding the problem of locating technological functionality is in no small part a question of understanding who will use it, and when. Educational activity for young children, we suggest, typically requires others to be involved and has consequences for any personalisation by location. Video evidence showed how young children will frequently play around the kitchen table while the adult works at some domestic task. On the other hand, when specifically educational work needed to be done, adult and children have to leave the kitchen in order to do so. It would appear that here is one obvious reason for having a computer in the kitchen.

Given that technologies are frequently designed with a single user in mind, or otherwise with groups of users in mind, this evidence would seem significant. It is, of course, important that a given design actually reflects the real-life group/individual dynamic it is designed for, and our evidence suggests that, in the domestic environment, a great deal of care is needed to distinguish one from the other, and more pertinently, on what occasions we see a preference for one or the other.

---

[19] Conversely, parents often seemed simply to presume that the computer activities of older children were educational, without any direct monitoring.

Older children are more likely to use PC functions, but again their use consistently orients to their social needs. In the case of one 12-year-old, the main uses for the PC, leaving aside games, had to do with visiting various websites. He spent some time explaining how his favourite while at the house had been Boltblue.com. His comment was as follows: "BOLT-BLUE – It's great 'cos you can contact your friends for nothing – e-mail or SMS. My stepsister showed me Boltblue, and I use it after I've finished my homework. They have thousands of icons, I've got one on my phone – they've got loads of categories like sport, music, cartoons. You're only allowed to download two a day – I've used up all my limit for the month. It has ring tones as well, but I can't download them because my phone doesn't have Composer. It's a 5110."

The same boy also commented very positively about the master bedroom screen, "We spent a lot of time watching movies on the big screen. It was cool. It was a bit like being in the cinema." Even so, we should not give the impression that this closeness was continuous. This 12-year-old also said, "I sometimes watched TV in the family room to get away from my sisters. I couldn't use the remote to switch over to Sky so I had to get up to switch over (and I like to flick). I escape here . . . watch TV on my own. I did play with the girls though. I drew pictures with Sammy and Em. Sammy and I would print out notes for each other and leave them on our beds."

Touchingly, and revealingly, he also commented, "yeah, I have used the computer a lot while I've been here. But I don't so much at home. The truth is I get a bit lonely and I miss my mates here. Like, I always do my homework with them at home . . . " Although relatively young, this boy's comments resonate with sociological research about teenagers and "bedroom culture" (see, for example, McRobbie, 1991), and one might venture to suggest that the applicability of technological functionality in this context should be understood in and through the peculiar and somewhat marginal status of the teenager.

12.3.2 Distance Connectivity

Observations led to the view that issues of distance connectivity are important in two ways. First, the direction of monitoring and information flow is important, and second, issues such as immediacy and image quality seem less important than simple sociality and the historical sense that wider networks provide. To begin with the issue of direction, there is a world of difference between being able to monitor the world outside the front door and the world outside the front door monitoring you. This became evident when the use of the health and medical monitoring facilities was observed. All three families reported much the same thing regarding the exercise and health facilities, summed up in the following comment: "We started off using it but it dwindled away. For a start, we're not as fit as we thought we were. I'm not a hypochondriac so I didn't really need the help. We didn't get any feedback from them in any case – perhaps that should be reassuring – the nurse was very thorough when she came round – if you actually had some condition it would be very good. The nurse suggested we did it every day, but we didn't . . . just occasionally. It wasn't really for us. To be honest, I just didn't like it . . . I don't like being constantly monitored . . . "As another woman said, "it's good. I like the idea of checking blood pressure, cholesterol etc. but only for my own consumption – not outsiders."

In contrast, information-seeking behaviour around health was not unusual: "I did ring them up one time – it's a 24 hour advice line. [my daughter] had a rash and it spread over the course of a night. She was whingeing and I was debating whether to take her to a doctor and they told us we should." The same kind of general enthusiasm for health information is to be found in the following comment: "I had a contraceptive injection and I browsed the net and had a look – there were thousands of women who had the same side effects as me. It was brilliant to be reassured. I did a search on the name of the drug, and found it posted on a bulletin board. My doctor never said anything about side effects like emotional and hormone problems. I could talk to other women who felt the same way as me in a chat room". In sum, where family members are unenthusiastic about outside monitoring, they are positive about the affordances of technology when the direction of the monitoring is outward.

A striking feature of all our families was their enthusiasm for any technology that allowed them to be connected more widely, especially to other family members and to special interest groups. Hence: "I talk to a lot of people about bike stuff. I sort of know these people. I like it. I just stumbled across 'Bikers Café'- I just found it. I like the people in the Café, it's a nice social scene. I've been using it for about two years. It's been very frustrating not to be able to do it here. Lee uses chat rooms as well, with other kids." (Any value in mobile access to chat rooms?) "Probably not . . . I mean, there's only so much time . . . teenagers might . . . "

Perhaps the most striking feature of the research was the universally positive reaction to the affordances of digital photography. Video data showed the extensive use to which families put the digital camera, the display screen, and the printer. Even if we factor in their natural desire to record as an "occasion" their visit to the "smart house", the delight in the affordances of digital images was apparent across all families, and more or less regardless of age. Thus, the 12-year-old boy in particular was a great user of this camera: "I haven't got a camera. The digital camera was easy, though I've only just found out you can print out all your photos at once. I didn't know how to save them to the computer, so I was sending the image straight from the camera to the printer." His mother commented: "He loved the digital camera. I think its fantastic as well. Have you seen all the photos he's printed out? [There is a large pile of printed images on the kitchen table] I think its great. My sister has one, she's already sent us a CD full of photos." The mother in another family, a self-confessed technological illiterate, was entirely positive about this. As she said, "I like taking photos, and I always have them developed in a 7 by 5 format. That's very expensive, and a waste of money if your photos are rubbish. The digital camera was just fantastic. It costs you more or less nothing to take photos, you can chuck away the rubbish ones without developing them, you can print them out cheap, and if you want you can buy high quality paper and print them out on that. That's what we did. We also found out there are firms that will print them for you (on the Net). We're going to buy one." She made a further point: "I already use e-mail, but the reason is because my sister's profoundly deaf. So she can't talk on the phone. This would be great for sharing – I could send her pictures all the time."

There has been some research on the role of photographs in family life, research which stresses the inherently social nature of the image (see Frohlich et al., 2002). Put simply, looking at pictures is something that is typically done as a group. Families will review recent experiences, share them with other kin or with friends, and use images as a focus for recall and discussion about these experiences. A significant element in this popularity is the way that digital images can be conveniently meshed with ordinary family concerns to record their history, and to relive significant occasions of family life. The popularity of Net meetings can very much be seen in these terms: "Net meeting would be a popular option with us. With the speed of the access here, and the *bandwidth*, that would be fantastic. Actually, the image quality isn't that important to us. I can tell enough. We can still see [our niece] growing up. Through Net meetings, our friendship networks have actually grown, like my sister now knows some of my other friends and will talk to them even when we're not logged on. With MSN you can send files more or less immediately, so you can look at photos and stuff like that."

Interestingly, these web conferences with wider family did not seem to depend on immediate interactional affordances, and the above comment may help explain why. Family interactions of this kind may well be more about getting historical markers for family relationships – the niece's size since they last saw her and such like – rather than the ability of digital imagery to convey gestural information. If so, this has some profound implications in terms of the difference between home and work settings, particularly in the context of video conferencing. Where video conferencing has hitherto been something of a niche market, largely predicated on observations concerning gesture, gaze, etc., it rather seems here as if these features are less important because there is seldom any immediate task at hand. There would appear to be considerable mileage in continuing to explore this theme, especially as Frohlich et al.'s research into photography in the home has shown how important this process of family maintenance seems to be. One avenue of exploration would be to compare teenagers' use of such devices with family-oriented use. A second has to do with display technology in the home, since two of the three families showed clear willingness to play with digital display, and particularly to identify how such display technology would be used on occasions such as "family get-togethers", especially when some family members are missing.

In many ways, these tentative results echo Frohlich et al.'s work. This describes the different types of conversation that take place on the telephone, notably single topic, purposeful calls and multi-topic calls which are more concerned with maintaining personal relationships rather than with the achieving of specified objectives. Their work points towards guidelines for the development of technologies to support the types of conversation that happen. In much the same way, if we are to develop domestic technologies to support distance connectivity, it must be through an understanding of what people actually do in these situations.

12.4 Conclusion

This chapter has argued for the inclusion of domestic life and the new technology that might be associated with it into the CSCW research programme. While entirely in agreement with Hindus (1999) regarding the fact that domestic life is substantially different from working life, there is one respect in which it is analytically the same. That is the way in which new technology in the home may have to be understood in terms of its interactional affordances. It is for this reason that the notions of control and social connectivity have been emphasised. In our view, a prior emphasis on location, while entirely valid, does not wholly come to terms with the rhythms of family life. Issues of control cannot be reduced to cognitive load. They include not only the individual's sense of being able to use the technology, but also the sense of control that comes from knowing what others are doing or have done with the same technology. It appears that significant feedback is necessary if that sense is to be maintained. Equally, families are not "units" in a behavioural sense. Neither are they collections of individuals who happen to live in the same place. They can be understood as individuals who orient to their family membership at specific times and in specific ways. Our data show how family members both avoid each other and seek their company, can be engaged in activities which entail them being alone, and otherwise act collaboratively. The point is, of course, to distinguish which is which, and when. That is, a sophisticated view of social connectivity will be necessary.

Social connectivity comes in two distinct forms, called in this chapter "local connectivity" and "distance connectivity". These refer to the quite ordinary respects in which family members orient first to one another and second to others outside the home. Both are interesting and important. The first is important because it pertains to, along with issues of control, the problem of personalisation and point solutions. Local connectivity, one might argue, is a critical issue for the desirability of personal, point solutions in the household. In the near future, however, other forms of personalisation will become more salient. The likely reason for this is the spread of networked devices through the home, just as the network has become the default in workplaces. The prospect of most computer-related devices in the home operating from one central server opens up a whole range of possibilities, of which the *personalisation of the interface* is one. If we again take the kitchen as an example, we can see that different screen sizes may be appropriate in different locations. (Screens near the fridge and/or cooker will not need to be as large as one at the kitchen table.) Given the problems of control, which were referred to above in terms of overhead, reliability and so on, the use of information resources will depend on how quickly and easily information can be input, used and retrieved at various locations, which will in turn depend on dedicated menu structures/local interfaces. For example, the use of lists in kitchens, and a variety of technologies suggested for use in association with the list (e.g. bar coding; automatic food ordering; prompts for suitable meals, and so on) will depend in part on the elegance and immediacy of the design solution in question.

In some respects, the kind of smart house referred to above as the "learning home" and "alert home" will deal with these issues. Thus, the kind of personalisation the learning home will deliver will deal with the issues of control observed in the bathroom, which are captured by these sentiments: "why can't you . . . why can't you specify a temperature for each person, and an amount to fill it up. That would be great, wouldn't it? I could just input [the name of son] and he'd get a lukewarm bath, which is what he likes. Me, I like it scalding."

Such an arrangement would obviate most complaints about lack of control in the bathroom, and indeed elsewhere. We might call this *personalisation by profile*. Nevertheless, this kind of personalisation is also fraught with difficulty. We have seen how problems arise with the control of security, entertainment and other systems in the home. Some of the reported difficulties have to do with simultaneous commands, lack of feedback, and the absence of a clear structure of priorities. Such systems require more than command structures that allow different individuals access to different menus. They also need a sensitivity to the history of control, covering such matters as who last used the device, when, and what for. This raises the issue of entitlement – who has a right to override other commands, and who has not.

A third personalisation issue is *personalisation by activity*. When kitchen equipment and the possibility of electronic support for shopping, cooking, etc., was discussed, few family members showed any interest. When they did, it was surprising how little they wanted. There was some support for keeping electronic lists using stylus entry on a wall-screen, along with prompts (presumably from the fridge and/or the cupboards) indicating that certain goods were running short. One mother, when asked if recipes on a screen would be useful, said, "not really . . . mind you, if they were connected up to the oven and the microwave, so they automatically went through the right heating sequences and the like . . . that'd be good . . . " What this indicates is that we need a much better sense of what the activities in question actually are before we can decide on the usefulness of technologies to support them.

Finally, and most profoundly, the issue of personalisation depends on the degree of local social connectivity observed, and this affects *personalisation by location* above all. The expected move away from the PC in the home will have to be accompanied by some careful consideration concerning which kinds of devices and applications will be appropriate in which location. It is clear, for instance, that we need to know a great deal more about the behaviour of young people vis-à-vis educational experiences in the home. As suggested, the PC is inappropriate as a bedroom-based resource for educational work, for the simple reason that educational work typically turns out to be collaborative. Given that parents are often busy with other activities when demands are made on their time, consideration must be given to the control surfaces for the kitchen. It is also relevant to patterns of entertainment use, for the location and type of entertainment systems will depend very much on the nature of family life. Several parents expressed anxiety about the way their children were more isolated, spent more time watching TV, etc. Increased personalisation of technology for children may well exacerbate that situation, at least among certain age groups. An area that needs appreciably more research is how personalised devices can be provided such that casual visitors to the home can also use them.

In this study, the issue of distance connectivity turned out to be the most surprising and most positive aspect of family life in the "smart house". It might be that we had taken for granted arguments about the primacy of small family units in industrial and post-industrial society. What was surprising was the sheer vibrancy of extended family connection, and the desire to expand it wherever possible. It is clear from these results that in many respects the extended family is alive and well. That is, regular contact with a widely dispersed set of family members should be regarded as a typical feature of modern family life, arguably more so with the advent of widespread communications technology such as e-mail. In a sense, of course, 'twas ever thus, given that the letter and telephone have existed for a long time. Even so, mobile telephony and text messaging, digital images and video, netmeetings and so on all afford regular contact with others, not only on an individual basis, but also collectively.[20]

---

[20] Any thoughts we might have about the way in which technology in the home might relate to younger people would benefit from more nuanced studies of teenage behaviour, because there was no opportunity to observe any such animal in this study. Having said that, it is apparent that connectivity in general, evidenced by the widespread take-up of mobile telephone and SMS messaging (see, for instance, Grinter and Palen, 2002) suggests that teenagers may well be a significant audience for some technological developments.

There is some evidence from this study that there is already take-up of these possibilities.[21] We might note a number of features of this distance connectivity, all of which are potentially important for the take-up of new technology in the home. First, where monitoring or information use is the issue, there appears to be a significant difference between outward-looking and inward-looking facilities.[22] The families in the study showed themselves to be uncomfortable with any facilities which they felt monitored their behaviour, even when it was for the best of reasons, such as health monitoring. This has to do with the obvious but often forgotten fact that family life is private life. A second feature has to do with the fact that issues of image quality and bandwidth did not appear to be very important for our families. This may be because the immediate interactions involving other family members or wider social networks are less important than the maintenance work involved in these communications. Digital imagery and video interaction in this context was above all a means to maintain family and social solidarity and history. It is perhaps for this reason that reaction to this kind of affordance was so positive. In turn, this presented the most surprising result of the enquiry. CSCW practitioners are familiar with the general failure of video conferencing to provide more than a niche market in organisational life. Tentatively, and bearing in mind the small nature of the sample, in the medium term it may well turn out to be much more central to domestic life. Indeed, one family in the sample was adamant that they had made new friends, and that their networks had spread and become more dense, as a result of their online activities. Of course, some of the issues entailed will turn out to be the same, certainly in terms of document or image sharing. Nevertheless, the increasing popularity of new forms of net meeting, for instance, is an eminently researchable arena.

In summary, it is clear that patterns of connectivity are difficult to predict on the basis of this small sample. Having said that, we can perhaps think in terms of two axes for a matrix model which might inform attention to domestic life studies in the future. On one axis is the closeness/distance continuum, and on the other are the various social factors that

---

[21] A significant feature of this, however, is that such kinship connection does not take place only at a distance. Again, a common way of expressing family values is through the ordinary rituals of life, including births, marriages, etc. Equally, Christmas and other festive occasions are also treated as occasions for family "get togethers". The existence of powerful family networks of this kind is another potential source of bias within our sample, since issues of personalisation and stability, ease of use and control ought to be investigated not only in the context of the immediate family experience, but also in the context of links with family and friendship networks. For example, personalised technologies have an obvious conflict with visitor use. The regular occurrence of family occasions where wider kin are habitually pressed into service presupposes that user-friendliness may have to take this into account.

[22] We have no space to discuss the intermediate forms of connectivity involved in integrated security and communications technologies. These include applications which link remote householders to those wanting to access a home such as trades people and delivery persons; alarm monitoring, and so on.

might influence the need for connectivity. These include the needs of families with young children, with teenagers and their work of "doing independence", with "empty nesters" and their desire to extend their social life, and with the so-called "silver surfers" and the construction and maintenance of family history. In any event, the scope for future research of this kind is enormous.

References

Barlow, J and Gann, D (1998) "A Changing Sense of Place: Are Integrated IT Systems Reshaping the Home?", paper presented to the Technological Futures, Urban Futures Conference, Durham, 23–24 April.

Berg, C (1994) "A Gendered Socio-technical Construction: The Smart House2, in C Cockburn and R Furst-Dilic (eds.), *Bringing Technology Home: Gender and Technology in a Changing Europe*, Buckingham: Open University Press.

Bose, C, Bereano, P and Malloy, M (1984) "Household Technology and the Social Construction of Housework", *Technology and Culture*, Vol. 25, pp. 53–82.

Cockburn, C (1997) "Domestic Technologies: Cinderella and the Engineers", *Women's Studies International Forum*, Vol. 20, No. 3, pp. 361–71.

Cowan, RS (1983) "More Work for Mother: The Ironies of Household Technology from the Open Hearth to the Microwave", New York: Basic Books.

Frohlich, D, Kuchinsky, A, Pering, C, Don, A and Arris, S (2002) "Requirements for Photoware", Proceedings of CSCW '02, 16–20 November, New Orleans: ACM Press.

Gann, D, Barlow, J and Venables, T (1999) *Digital Futures: Making Homes Smarter*, Coventry: Chartered Institute of Housing.

Gann, D, Iwashita, S, Barlow, J and Mandeville, L (1995) *Housing and Home Automation for the Elderly and Disabled*, SPRU and the Electrical Contractors Association.

Gilbreth, LM (1927) *The Home-Maker and Her Job*, New York: D Appleton.

Grinter, RE and Palen, L (2002) "Instant Messaging in Teen Life", *Proceedings of CSCW '02, 16–20 November*, New Orleans: ACM Press.

Hardyment, C (1988) *From Mangle to Microwave: The Mechanisation of Household Work*, Oxford: Polity Press.

Harper, R, Randall, D and Rouncefield, M (2000) *Retail Financial Services: An Ethnographic Perspective*, London: Routledge

Hemmings, T, Clarke, K, Francis, D, Marr, L and Randall, D (2001) "Situated Knowledge and Virtual Education", in I Hutchby and J Moran-Ellis (eds.), *Children, Technology and Culture: The Impacts of Technologies in Children's Everyday Lives*, London: Routledge.

Hindus, D (1999) "The Importance of Homes in Technology Research", *Co-operative Buildings Lecture Notes in Computer Science*, Vol. 1670, pp. 199–207.

Hughes, J, Randall, D and Shapiro, D (1992) "Faltering from Ethnography to Design", in J Turner and R Kraut (eds.), *Proceedings of CSCW '92*, New York: ACM Press.

Kidd, CD, Abowd, GD, Atkeson, CG, Essa, IA, MacIntyre, B, Mynatt, E and Starner, TE (1999) "The Aware Home: A Living Laboratory for Ubiquitous Computing Research", in N Streiz, S Konomi and H-J Burkhardt (eds.), *Cooperative Buildings: Integrating Information, Organization and Architecture*, Proceedings of CoBuild'98. LNCS 1370. Springer, pp. 190–97.

Laslett, P (1972) "Mean Household Size in England Since the 16th Century", in P Laslett, *Household and Family in Past Times*, Cambridge: Cambridge University Press.

Martin, D and Bowers, J (1999) "Informing Collaborative Information Visualisation through an Ethnography of Ambulance Control", in S Bodker, M Kyng and K Schmidt (eds.), *Proceedings of the Sixth European Conference on Computer Supported Cooperative Work*, Copenhagen, Denmark: Kluwer.

Mateas, M, Salvador, T, Scholtz, J and Sorensen, D (1996) "Engineering Ethnography in the Home", *CHI 96 Electronic Proceedings*, Vancouver, BC: ACM Press.

McRobbie, A (ed.) (1991) *Feminism and Youth Culture*, London: Macmillan.

Mozer, MC (1998) "The Neural Network House: An Environment that Adapts to its Inhabitants", in M Coen (ed.), *Proceedings of the American Association for Artificial Intelligence Spring Symposium,* pp. 100–14, Menlo Park, CA: AAAI Press.

O'Brien, J, Hughes, J, Ackerman, M and Hindus, M (1996) "Workshop on Extending CSCW into Domestic Environments", in *Proceedings of CSCW '96*, November 1996, p. 1.

Schmidt, K (1991) "Riding a Tiger, or Computer Supported Cooperative Work", *Proceedings from the 2nd European Conference on CSCW*, Dordrecht: Kluwer.

Scholtz, J, Mateas, M, Salvador, T and Sorensen, D (1996) "SIG on User Requirements Analysis for the Home", in *Proceedings of CHI '96 conference companion*, p. 326.

Vanek, J (1978) "Household Technology and Social Status", *Technology and Culture*, Vol. 19, pp. 361–75.

Venkatesh, A (1996) "Computers and Other Interactive technologies for the Home", *Communications of the ACM*, December, Vol. 39, No. 12, pp. 47–54.

Wajcman, J (1991) *Feminism Confronts Technology*, Cambridge: Polity Press.

Smart Home, Dumb Suppliers? The Future of Smart Homes Markets

13

James Barlow and Tim Venables

> Mitch was bored with being Ray Richardson's technical coordinator. He wanted to go back to being an architect, pure and simple. He wanted to design a house, or a school, or maybe a library. Nothing showy, nothing complicated, just attractive buildings that people would like looking at as much as being inside them. One thing was for sure. He had had quite enough of intelligent buildings. There was just too much to organize (Kerr, 1996, p. 43).

13.1 Introduction

Thirty years ago, Nicholas Johnson argued that the home would ultimately become a

> home communication center where a person works, learns, and is entertained, and contributes to society by way of communications techniques we have not yet imagined – incidentally solving commuter traffic jams and much of their air pollution problems in the process (Johnson, 1967, quoted in Graham and Marvin, 1996, p. 92).

Numerous writers and filmmakers have speculated about future homes, sometimes in threatening, sometimes in comical terms. In many of these visions, the home is seen as a physical access node for "electronic spaces" within advanced communications networks. Typical is Alvin Toffler's notion of the "electronic cottage" as a locus for employment, production, leisure and consumption [23]. Common to many perspectives is a redefinition of the home to allow the household to reassume roles – such as work, education, medical care and entertainment – which have increasingly been externalised. According to Toffler, a desire by individuals to retreat from the environmental, social and political problems of modern industrial cities lies behind this trend.

The evolution towards the multi-functional home is also, some suggest, a result of changes in the spatial organisation of advanced capitalist

---

[23] Others have written on the "computer home" (Mason and Jennings, 1983), the "electronic house" (Mason, 1983), and the "smart home" (Moran, 1993).

society. As Lorente puts it, "global houses" are needed if we are to have "global villages". He feels that fully inter-connected housing can act as an interface between Castells' "flow space" – the increasingly important network of information flows – and physical space, where the experience and daily life of most people takes place. In this way the home can become part of a world of dialogue between people, between people and machines and between machines themselves. The home will not only be a passive receiver, but also an active producer of information and energy, the latter through the generation of solar electricity.

Elements of all these observations can be observed in contemporary society, with important implications for the way people live. However, despite the projections, we have yet to see them fundamentally affecting *the homes* in which they live. There has been only limited progress towards the introduction of "smart home" technologies. The view, expressed in 1989, that "a combination of home computers, consumer electrical goods, videotex services, and home security systems, even in a "smart house', wired with heating and lighting sensors . . . hardly add up to a revolution in ways of living" (Forester, 1989, p. 224) still largely holds true.

Nevertheless, change *is* occurring, as new communications and information technologies become "domesticated" . The aim of this chapter is to review progress towards the "smart home" and outline perspectives on the way technology change is beginning to accelerate this process. The chapter suggests that past approaches by industry and government, based around the demonstration of fully integrated smart homes,[24] have failed to generate sufficient interest from consumers for a mass market to form. However, a change in focus can now be observed, with new industry players entering the market. These are beginning to address potential user needs that have emerged in conjunction with the development of information and communication technology (ICT). On their own, networks and services outside the home will not, however, be sufficient to ensure universal access to the "information society". In order to take full advantage of new informational services, it will be necessary to distribute and make them accessible once they have crossed the threshold of the home. This is leading to new interest in aspects of smart home technologies, as well as a revision of the concept of the smart home.

13.2 "Smart Homes": A Recent History

The terms "smart homes", "intelligent homes" and "home automation" are often used interchangeably. In essence, they embrace two approaches to classifying the technologies that shape the use of the home:

[24] For example Japan's TRON (The Realtime Operating-system Nucleus) Project in the mid-1980s, Legrand's Domotique system in the early 1990s, National Panasonic's Home Information Infrastructure (HII) house in the late 1990s and Orange at Home in 2001.

- First, the notion of the smart or intelligent house captures the idea that the material environment of the home and domestic tasks can somehow be automated. Automation can range from (1) simple fixed applications with predefined and pre-established operations, through (2) programmable applications and devices to (3) fully flexible and automated applications and networks of devices that share information and provide it to consumers. From a functional perspective, automation can be designed for *convenience*. Central locking of doors and windows, lights coming on when someone enters a room or telephoning the house to start the bath running are all examples. As has been shown in previous work (Gann et al., 1999), rather than merely offering a marginal increase in convenience, this type of functionality can be of major benefit to certain population groups, such as older or disabled people. Some of the same applications – such as control over lighting – can also have a functionality more related to *building management and environmental* control. Domestic energy management systems and comfort control, and fault diagnostics would fit into this category.

- For most people, these smart home technologies offer additional convenience in everyday activities, incrementally adding to the benefits provided by previous mechanically and electrically based eras of technological change in the home (Gann et al., 1999). However, the emergence of digitally based ICTs is now offering the potential to greatly enhance the functionality of the home by providing an interactive window to the world outside, and by providing us with information and feedback that was previously impossible to obtain. The second broad area of smart home technologies therefore stresses the notion of the "informational" home, where existing and new information services are used to improve the management of family and professional life. These services can range from the improved distribution of existing prevalent electronic communications (analogue TV, telephony) to new electronic services (broadband Internet access, digital TV). Some information-based services are provided on a non-customised basis as part of general entertainment or educational services, either "pushed" by the service provider or "pulled" by the customer. Developments in ICT and the emergence of new service providers are, however, also creating the possibility of more customised services, provided direct to individuals or to the homes in which they live.

Developments in these two areas are leading to the possibility of greater integration of household functions within homes, and between homes and externally provided services. Combined in the right way, they may achieve the goal of increasing functionality in the home. However, this will only occur if previous models for the introduction of the smart home technologies are avoided. It has traditionally been held that the ability to communicate and control lies at the heart of the smart home, and that it is necessary to integrate systems in order to provide the types of functions that people will want to use. Many demonstration projects have followed this model and involved major capital investment to deliver a utopian vision of the "home of the future". These generally necessitate a fundamental change in occupants' day-to-day way of living.

While these homes have stimulated interest, they have not provided consumers with a realistic model that can be implemented in an affordable manner. This is partly due to suppliers pursuing an approach that is largely technology-push rather than demand-pull. Technologies and components developed for use in commercial and industrial buildings have only been marginally modified for domestic application (Gann et al., 1999). While some products specifically designed for the home have been developed, there remains a gap between consumers' requirements for systems which are useful for managing everyday tasks and the available products. Furthermore, the integrated nature of many systems and products requires specialist knowledge and/or training before they can be installed and used.

13.3 New Players and New Markets?

There are a number of reasons for believing that the picture of market failure described above may now be changing:

- the role of major ICT players in promoting smart home products and systems is changing;
- ownership of personal computers is continuing to rise;
- voice, data and image (VDI) media are increasingly digitised;
- e-commerce is leading to the development of services for the residential market.

13.3.1 The Role of Major ICT Players

In the 1980s and 1990s, firms such as Siemens and Legrand were adapting intelligent buildings technologies originally developed for the office market for the home, with limited impact. At the same time, other manufacturers, predominantly in the telecommunications and consumer electronics markets, were developing new digital systems and components with domestic applications. Key developments include the replacement of electromechanical switching by digital switching and traditional twisted pair and coaxial cables by optical fibres. Other trends potentially supporting smarter homes also emerged during this period. These included new communication networks (ISDN, xDSL, the Internet), which allow bi-directionality (two-way communications), and developments in end devices such as web TV and video telephony.

These trends led to companies, including Sharp, Sony and Microsoft, introducing new products and services aimed at the home, based on these emerging technologies. Some consumer electronics companies now believe that the model for introducing smart home devices needs to move from one in which a number of products and services are provided as a fully integrated package, towards one in which customers install network-ready devices in an incremental manner. This view is based on the premise that consumers are unlikely to commit themselves to the likely major investment and disruption involved in the installation of a fully integrated system. Rather, consumers may wish to move towards their ideal of a smart home by purchasing "smart" or network-ready devices to be integrated at a later date.

A new set of domestic products is therefore emerging which allows this possibility. For example, kitchen appliances by Electrolux, Ariston and LG can be connected to a smart home network or a telephone system, enabling the ability to download recipe or software updates, monitor performance, alert service centres of impending faults or engage in e-commerce. A variant of this approach is the incremental purchase of devices that can communicate with each other, as specific needs arise. Sony already provides a high level of interoperability between many of its consumer electronics products – cameras and computers can be connected using iLink, and the "Memory Stick" product allows easy transfer of data between computers, music and video systems, and their robot pet series. Sony is now intending to expand this model to their other products.

The incremental approach therefore supports the installation of devices according to a consumer's perception of need or because an existing product has reached the end of its lifecycle. It is also especially suitable for the existing building stock, where refurbishment tends to occur on a system-by-system (e.g. central heating) or product-by-product basis, with people replacing existing consumer durables with ones that are future-proofed for emerging technologies and services when they need to. The longevity of the housing stock and low rate of replacement in many countries means the mass market for smart homes products or services is therefore likely to be in refurbishment or installation in existing properties, rather than new housebuilding.

13.3.2 An Increase in Domestic PC Ownership and Internet Access

In 1999, PC ownership among households in European OECD countries ranged from 65 per cent in the Netherlands to 21 per cent in Italy. Internet access from the home was generally lower, but is growing faster because the prior investment in PC equipment and skills has already been made. The rapid increase in PC ownership over the last decade – and more recently, ownership of multiple computers in the home – is stimulating demand for dedicated computer networks. By this we are referring to wired or wireless local area networks (LAN) used to share resources between computers and peripherals. Providing LANs with access to wider networks, perhaps through some form of broadband connection, introduces the potential for home servers which store more than conventional data – notably audio and video material – and act as an in-home web and mail server.

13.3.3 Digitisation of Media

Since the early 1980s, there has been a steady shift from the use of analogue technologies for the transmission and storage of information towards digital technologies. This was first manifested in telephony – fixed line and later mobile – and is now spreading to other areas, notably TV transmission. More recently this trend has extended to consumer electronics products, such as cameras and audio systems.

With regard to media transmitted to the home, digitisation has started to break down the distinction between voice, data and image (VDI), making it possible to converge distribution methods around the home. This obviates the need for separate and distinct networks for every application and facilitates economies of scope in distribution. For example, it is already possible to combine telephony and computer data transfer over the same network. The key limitation to this is one of bandwidth. The existing telephone network will transfer computer data, but at a low speed, while a dedicated computer network could be used for multiple voice transmission. While digital transmission notionally reduces the bandwidth required for the transmission of a signal, in practice it offers the prospect of additional services utilising any spare bandwidth. A good example is digital TV, where sports channels are able to offer surround sound, widescreen display and multiple parallel channels offering different perspectives on the same sporting event.

Digitisation is also affecting the storage of media. Within the last few years, photography, music and video – and now print media – have all started to be distributed and enjoyed in new electronic formats, rather than through the exchange of traditional physical artefacts. Emailing a photo to family members is no longer considered exceptional and the music industry is currently undergoing a revolution in the way music is purchased and listened to. These trends make it necessary for the introduction of appropriate systems for importing data from outside the home, distributing it around the home and storing it securely within the home.

13.3.4 E-commerce and the Emergence of Services to the Home

A wide range of new services has emerged in recent years, through the penetration of the Internet. Most notable has been its use for e-commerce, especially shopping and banking. There have been spectacular commercial failures, but there is confidence in both industry and government that e-commerce will become a critical aspect of business in the future. Moreover, use of digital TV has expanded rapidly since the late 1990s, partly as a result of demand for additional channels and improved broadcast standards. This, along with developments in mobile telephony, is likely to stimulate access to Internet among groups that do not own a personal computer, as has been the case in Japan with the introduction of the I-mode mobile phone technology.

As well as home or mobile shopping, e-commerce is being stimulated by the electronic delivery of government services. In the UK, there are moves to use the Internet and free-to-air digital TV to improve access to information on health, housing, education and local and central government services (Cabinet Office, 2000a, b; DETR, 2000). And a range of other suppliers – notably telecommunication companies, energy and water suppliers, and health and social care authorities – are developing systems to allow greater connectivity and the provision of value-added services. These include interactive and multimedia information and entertainment services, remote energy management and automated monitoring and control of domestic appliances, and telecare.

13.4 Networks and Transmission Technologies

The parallel development of these ICTs has led to the emergence of a tranche of players interested in developing smart homes markets. While these are predominantly focusing on information distribution rather than home automation, the technologies they have developed demonstrate plenty of scope for interoperability with home automation systems. This will be further enhanced by the continuing development of new network and transmission technologies. However, the services described above also require distribution networks within the home and the ability to link these networks to the outside world. Moreover, there may be a requirement for multiple internal networks, dependent on the desired applications, ranging from a narrow band network for home automation purposes, through medium band networks for sharing computer data, to a broadband network for the distribution of audio-visual content. The key technical challenge is to develop appropriate standards and protocols for interconnectivity between these networks.

There are currently a number of standards and protocols for internal networks (see Table 13.1), but no single standard is appropriate for all purposes. There is, therefore, a need for the development of suitable middleware to absorb the differences between protocols, allowing various networks to inter-communicate with each other. In addition, in-home networks will require a means of connecting with wider area networks. This is not essential for home automation functions, although it may be desirable for remote control purposes and for facilitating access to wider services. Currently there are a number of wide area networks (the

Table 13.1. Examples of internal network types

	Transmission media	Protocol
Narrow	Powerline	Konnex
	Dedicated bus	LonWorks
Medium	Wired LAN	Ethernet
	Wireless LAN	Home RF (IEEE 802.11b)
Broad	Wired – copper and fibre	IEEE 1394
		xDSL
		Cable and digital TV protocols

Internet, fixed and mobile telecommunications, terrestrial, cable and satellite TV) that are linked to homes. However, these are usually delivered to a single point within the home (phone, modem, set-top box), with limits on their onward distribution. In order to achieve a true "informational home", it is therefore necessary to link these services to local networks in the home in an interoperable manner. This will require changes to the terms and conditions under which some of these services are provided – satellite or cable TV licence users are only able to watch a single channel per subscription. Interoperability may also be assisted by the development of open specifications, such as those proposed by the Open Standards Gateway Initiative (OSGI).

The link between service access distribution networks and in-home networks and devices is often referred to as a "residential gateway". This needs to be capable of mediating between the current and future varieties of transmission media and protocols to allow for future technical change. The residential gateway should enable service providers and application vendors to offer a variety of services without consumer intervention. It has also been suggested that such a gateway would need to act as a firewall between the internal and external networks to maintain the security of the home. Gateways can either be centralised within a single box or distributed across a number of boxes in physically separate locations which provide optimised entry for the various external connections. Gateways are likely to be modular, with open-ended architecture that allows appropriate selection of the plug-in interfaces for the desired services.

As well as the development of fixed networks both within and outside the home, and the options for interconnectivity made possible via gateways, it is also necessary to consider the impact of mobile telecommunications. This not only involves the various generations of mobile telephony, but also new channels for access and distribution, such as short-range wireless (e.g. Bluetooth). Mobile telephony, in particular, is rapidly becoming omnipresent and new generations of technology are enabling the provision of new services directly to individuals, regardless of location.

The evolution of these technologies since the 1980s has been accompanied by an expansion of potential services. Figure 13.1 illustrates this process for different generations of mobile telecommunications technology. The shift from analogue to digital telecommunications (1G to 2G) provided greater bandwidth at better quality, facilitating the use of mobile networks for data transmission as well as voice communication. Nevertheless, 2G and its variants essentially provide services in the same manner as 1G. The emergence of 3G not only represents a new system, but a new model of service delivery. This opens the way to allow mobile telephony to move from a mono- to multi-service model, mono- to multimedia, and from person-to-person communication to person-to-machine communication. Further developments in the use of embedded radio technologies will eventually lead to an expansion of machine-to-machine communication in localised networks.

These trends have important implications for home automation. The emergence of short-range radio-based communications systems will facilitate machine-to-machine communication and offer increased possibilities for monitoring and control. When embedded in homes, vehicles, personal belongings and public environments, short-range radio potentially allows greatly increased interaction between devices and between devices and individuals, regardless of location.

Together, the result of these trends is that the distinction between the home and the outside world is blurring. Services such as telecare, information or entertainment, which could previously only be provided via home-based networks and devices, can now also be provided to individuals whenever and wherever they need them. Equally, the operation of functions within the home is no longer dependent on physical presence within the home.

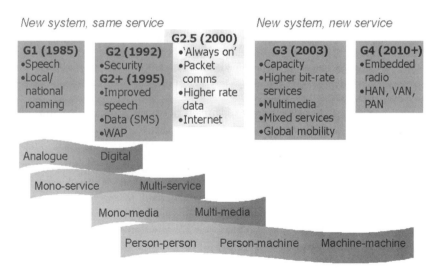

Figure 13.1 Changing technologies and service models in mobile telephony.

13.5 Models for the Future

We have argued that an approach to the introduction of smart home technologies that requires full integration of devices has proved inadequate for generating widespread consumer interest. Nevertheless, certain elements of this model need to be retained. The ability for components and systems to intercommunicate remains at the core of any smart home, but it is becoming increasingly apparent that this does not, and in some cases cannot, involve exclusive use of a single communications medium and protocol. There therefore needs to be a more sophisticated approach to understanding the processes by which a mass market in smart homes products and services operates. Table 13.2 illustrates five models of in-home networks, ranging from the most basic to a network of interoperable systems. The characteristics of each model are expanded on below.

Model 1 involves simple on/off switching systems for selected applications and requiring no additional network installation (e.g. remote control switching). This cannot be classified as smart home technology because there is no intercommunication between systems. In contrast, under model 2, smart appliances form island systems based on selected applications. These appliances tend to comprise white goods – predominantly kitchen appliances – featuring some level of interconnectivity and allowing functions such as workload monitoring, inventory control, downloading information (e.g. recipes) and interactive messaging for the home. Sharp, for example, is moving towards ensuring all its white goods are network-ready – able to send and receive information, albeit without a specific network protocol being defined.

Table 13.2. Models of home networks

Model	Relationship	Name	Characteristic
1	1:1	Remote control	Stand alone device-to-device relationship.
2	1:n	Smart appliance	Ability to send and receive information to and from a remote system.
3	n:1	Smart system	Ability to share information around components in the same subsystem.
4	n:n	Network of systems	Ability to share information between similar systems using the same protocol. Generally installed as a single event.
5	N:N	Intercommunicating systems	Ability to share information between diverse systems using differing protocols as appropriate to application. Probably installed in an incremental manner.

Notes: n = a limited system or set of systems; N = a large set of systems.

Model 3 comprises small, dedicated networks such as whole house lighting systems or improved control over central heating or home entertainment. These use the same technologies, but for a limited set of functions within a single subsystem of the home. They can, however, be expanded to include additional subsystems of a similar nature (e.g. central heating could link into an automation system for windows). Under model 4, there is a network of systems where all devices and subsystems communicate with each other, using a unified protocol. This is the traditional approach to smart homes. Finally, model 5 comprises interoperable networks – the merging of the automation and entertainment domains of smart homes through use of middleware.

Firms involved in the development of smart homes initially tried to move from simple one-to-one systems to much more complex $n:n$ systems. This approach was too large a step for both consumers and intermediaries involved in the sale and installation of smart home technology. A more appropriate development cycle would initially focus on $1:n$ and $n:1$ relationships, where consumers purchase smart or network ready appliances or install networked devices within specific systems in the home to address specific needs. This would form the basis of a modular approach to smart home development. As appliances and limited networks become more prevalent, demand may well then shift to more integration between devices and networks, eventually leading to the $N:N$ model. Such a model would evolve over time within a home, with components and subsystems being introduced to address the changing needs of the consumer.

13.6 Future Challenges

The potential market for smart home products and services is now looking more hopeful than at any time in the past. Consumers are starting to recognise their own needs and industry is starting to take a more considered and mature view of how these needs could be addressed. We have argued that this is the result of major ICT players developing a more sophisticated perspective on the market, as well as the outcomes of background trends such as the rise in the ownership of personal computers, the digitisation of VDI and the emergence of e-commerce. In the short term, the underlying technologies for supporting the smart home are developing rapidly and new solutions to some of the technical issues are being found.

The size of the future market for smart homes is, however, unclear. Despite technological progress and more favourable structural conditions, the current state of the smart homes market in much of Europe can be described as one of "unconscious inactivity" (see Table 13.3). Although products and services are being developed, industry has yet to signal the benefits to the general user. As yet, there are few signs that the market has moved towards the next stage and bottlenecks to the expansion of the market remain. These primarily relate to organisation issues and consumer attitudes.

13.6.1 Organisation Issues

From the consumer perspective, it is essential that there is a coherent framework within which smart home systems can be procured and implemented. It is unclear whether there will be a need for specialists to install and/or commission smart home systems. While currently complex, there are moves towards more consumer-friendly systems, which can either be installed on a do-it-yourself (DIY) basis or by a conventional electrician and remotely configured by a service provider. The problem is that there is currently no single smart homes "industry". In the past, system developers and installers have been small and technology-driven firms, unable to open up consumer channels because they did not have the necessary skills for marketing the concept of the "smart home".

New players are, however, emerging, drawn from major consumer electronics, TV and telecom companies. This offers hope for a more coherent industry in the medium term, but the way in which the smart home supply chains could emerge is still unclear. How the current – or any future – players will be involved in installation, integration, service provision and possibly service aggregation remains to be seen.

The business models for providing services both to the home and to the individual are also in a state of flux due to changes in technological possibilities. There are many potential players in this field, such as fixed and mobile telecommunications operators, television companies, utilities and shopping and banking services. The distinctions between content providers, service providers and distribution companies are evolving rapidly, but there are, as yet, no clear business models for services. In the short term, the structure of this industry is likely to remain unclear. Large

Table 13.3. Market development model

	Primary customer state	Market state	Typical activity required
Stage 1	Unconscious inactivity	No market	Product development, market research, education, standards
Stage 2	Conscious inactivity	Emerging market	Demonstrations, measurement of benefits, dissemination, supply chain development, education, training
Stage 3	Conscious activity	Growth market	Quality control, market support, training
Stage 4	Unconscious activity	Established market	Consolidation, refinement, monitoring, challenge conventional wisdom

telecommunications companies and utilities may play a more important role, particularly given their need to generate new business opportunities following their heavy investment in infrastructure and licences. It is possible that these service providers may well provide a link to systems integrators capable of configuring smart homes systems and may even provide essential equipment, such as the residential gateway, on a lease basis.

For organisations involved in the development, supply and installation of networks and devices for the home, this picture of an uncertain market presents challenges. Assessing which technologies are most appropriate to invest in is especially difficult, given the lack of clarity over the way these technologies will be applied by the user. What is clear at present is that the emergent requirement for networks in the home either requires firms to develop new skills or new firms to emerge. While consumers could install some basic smart home devices on a DIY basis, more complex applications require specialist installation and integration within wider networks. The development of true "plug and play" smart home devices may alleviate the need for specialist installers, but manufacturers have yet to achieve this goal.

13.6.2 Consumer Attitudes

The smart home industry needs to demonstrate that it can provide solutions that satisfy real user needs if it is going to motivate consumers to buy its products and services. The added value of the smart home needs to be clear if a mass market is to emerge. Smart home solutions to real consumer needs have to operate at three levels (Gann et al., 1999): (1) as generic technologies, providing the basic, standard compatible building blocks for (2) context-specific systems, adaptable to a wide variety of dwelling types, and (3) personalised systems, tailored to specific individual and household requirements. Furthermore, solutions must satisfy a number of conditions:

- functionality – the equipment/system must have clear and unambiguous functions;
- ease of use – clear and simple user interfaces, interactivity and connectivity;
- affordability – for individuals and other housing stakeholders such as landlords;
- reliability, maintainability and servicing costs must be acceptable;
- flexibility, adaptability and upgradeability – systems need to develop as user needs change;
- replicability and ease of installation – systems need to be available as a standard, reproducible product;
- standards compatibility across applications and when upgrading within specific applications.

Evidence from the early days of the personal computer industry suggests that unless these conditions are addressed, the initial adoption of smart home technologies may be slowed.

13.7 Conclusions

In his book *City of Bits,* William Mitchell (1995) observed that:

> Once you break the bounds of your bag of skin . . . you will also begin to blend into the architecture. In other words, some of your electronic organs may be built into the surroundings . . . So "inhabitation" will take on a new meaning – one that has less to do with parking your bones in architecturally defined space and more with connecting your nervous system to nearby electronic organs. Your room and your home will become part of you and you will become part of them.

While this may be somewhat overstated, it is undeniable that the late 20th and early 21st centuries have seen accelerating moves towards an interconnected society, with the creation of work and personal virtual networks and greatly increased flows of information and data. The advent of the "information society" has brought major changes in the ways in which we live. Our homes have not seen a transformation that matches this pace of change and previous attempts to develop a mass market for smart homes have failed. However, the need for appropriate technologies to allow users to participate in the information society is providing a fresh stimulus for the smart home, as well as a redefinition of the concept. The combination of new communication networks and technologies, developments in end devices and appliances, and the growth of new electronic services for consumers is creating a space within which a new version of the smart home is emerging. This does not involve the full integration of devices and systems. Rather, it comprises a set of smart appliances and subsystems, able to send, receive and share information to and from remote systems. Within a home, these are installed on an incremental and modular basis, as and when the consumer's needs change or equipment needs replacing.

Fulfilling the consumer's needs remains paramount, though. While the potential users of smart home technologies have similar basic needs from their homes, the context in which these are expressed varies widely. Homes are diverse in their physical design, layout and age, and people's lifestyles vary considerably as a result of age, socio-economic and other circumstances. The value-added to the consumer is a smart home's perceived benefit, not its inherent "smartness" or "intelligence". Consumers have basic needs revolving around convenience, simplification of tasks and safety and security. Bringing the concept of smart homes to the mass market will require suppliers to clearly address these needs and demonstrate the additional functional and subjective benefits that the various technologies can deliver. A modular approach, where initial basic needs are addressed and additional functionality is then added to suit changing requirements and expectations, offers the best prospect for future market development. Such an approach is now being turned into business models by some firms.

There are, therefore, reasons for believing that the smart homes market has the potential to become a major growth area over the next decade. This will require certain key barriers to be overcome, not least the lack of a coherent smart homes "industry". A number of players are beginning to position themselves to form the nucleus of such an industry. These include consumer electronics, telecommunications and energy companies. Once clearer business models for providing services both to the home and to the individual emerge, the need for suitable access and distribution systems within the home will grow in importance. From a technical perspective, the main challenge is to develop appropriate standards and protocols for interconnectivity between multiple networks, both within and outside the home.

As well as increasing the functionality of homes, the smart home could also play an important part in facilitating access to the information society. However, on their own, in-home communications infrastructures will not bridge the "digital divide". External networks that are appropriate in both bandwidth provision and affordability are a fundamental requirement. Telecoms regulators, governments and industry will all need to play a role in helping to make local broadband connectivity cheaper and easier for all to access.

Acknowledgement

This chapter partly draws on research carried out under an EPSRC-supported project on telecare.

References

Cabinet Office (2000a) *E-government: Electronic Government Services for the 21st Century,* London: Performance and Innovation Unit of the Cabinet Office.

Cabinet Office (2000b) *E-government: A Strategic Framework for Public Services in the Information Age,* London: Cabinet Office.

Castells, M (1989) *The Informational City: Information Technology, Economic Restructuring and the Urban-Regional Process,* Oxford: Blackwell.

DETR (2000) *E-government. Local Targets for Electronic Service Delivery,* London: Central-Local Liaison Group of the DETR.

Forester, T (1989) "The Myth of the Electronic Cottage", in T Forester (ed.), *Computers in the Human Context: Information Technology, Productivity and People,* pp. 213–227, Oxford: Blackwell.

Gann, D, Barlow, J and Venables, T (1999) *Digital Futures: Making Homes Smarter,* Coventry/York: Chartered Institute of Housing and Joseph Rowntree Foundation.

Graham, S and Marvin, S (1996) *Telecommunications and the City: Electronic Spaces, Urban Places,* London: Routledge.

Johnson, N (1967) "Communications", *Science Journal,* October.

Kerr, P (1996) *Gridiron,* London: Vintage.

Lorente, S (1996) *The Global House: New User Opportunities in Automation and Information,* Madrid: Universidad Politecnica de Madrid.

Mason, R (1983) *Xanadu,* New York: Acropolis Books.

Mason, R and Jennings, L (1983) "The Computer Home: Will Tomorrow's Housing Come Alive?", *The Futurist,* Vol. 16, No. 1, p. 35.

Mitchell, WJ (1995) *City of Bits: Space, Place and the Infobahn,* Cambridge, MA: MIT Press.

Moran, R (1993) *The Electronic Home: Social and Spatial Aspects,* Dublin: European Foundation for the Improvement of Living and Working Conditions.

Pragnell, M, Spence, L and Moore, R (2000) *The Market Potential for Smart Homes,* York: York Publishing Services.

Silverstone, R (1994) "Domesticating the Revolution: Information and Communications Technologies and Everyday Life", in R Mansell (ed.), *Management of Information and Communication Technologies,* London: ASLIB.

Toffler, A (1981) *The Third Wave,* New York: Morrow.

Index

A

Active counters, 25
Active tables, 25
Air based networks, 4
Affordances, 10, 44–5, 51, 171, 227–8, 235, 239, 241
  definition of, 45, 228
  of banking, 60–1
  of digital images, 239–40
  of paper, 102–12
  of paper-mail, 101–14, 121–125
Anthropologically strange, 9
Artificial intelligence, 6
Assistive technologies, 10, 163–177

B

Banks, banking, 52–55
  Internet banking, 52–55
  Online, 52–55
Broadband Institute, 25
Brown goods, definition of, 33, 34
B2C, 11

C

Cognitive domains, 6
  loads, 121, 124, 174, 222
Cold media, 82
Computer Supported Cooperative Work, 6, 131, 169, 186–187, 227–8, 241
Conspicuous consumption, 80, 88
Context specific systems, definition of, 23, 259
Cultural probes, 171
  definition of, 169,

D

Digital newspapers, 208, 217–219, 222
Disability, theories of, 165–6
Distance, concepts of
  intimate, 133
  personal, 133
  public, 133
  social, 133
Durable complementary assets, 68

E

Electronic cottage, 247
Electronic Bills and Payment Presentment (EBPP) 101, 109
Electronic programme guides (EGPs), 9, 115–25
Equator project, 165
Ethnography, 41–61, 79, 94–5, 104–112, 184, 227–8, 235, 239, 241
  sociological, 6–7
Ethnomethodology, 6, 168, 183–204, 227

F

Family, concepts of, 229
  routines, 140, 142
Feminist analysis, 6, 25–6, 228–9
Front stage and back stage, 92–3

G

Generic markets, 28
  technologies, definition of, 23 259
GPRS, definition, 84
GPS, 173
GSM, 173

H

Hawthorne effect, 110
HCI, 6, 184, 200
HomeNet project, 136–7, 141, 145, 149
Home workers, homework, 11–12
Hot media, 82
Households, conceptual frameworks of, 32–33
  definitions of, 32

I

Infectious disease model, 69
Interaction mode, 3
Internet, 5, 8, 10, 201, 34, 48–60, 63, 65–9, 76, 79, 82, 84–9, 101, 2 108, 117, 127–8, 131, 207, 209, 212–19, 222, 231, 250–1, 253–4
  domesticating, 128
  shopping, 51, 56–8, 231–2
Interactive TV, 9, 115–125, 79–95

ISDN, 21, 251
Invisible-in-use, 199, 201

J

Jetsend, Hewlett-Packard, 5

L

Labour, concept of, 64
Lifestyle magazines, 22

M

Martini solutions, definition, 45
Medi-packs, 173–174
Medical nook, 25
Memory sticks, 251
Moral economy, 42, 43
  order, 7
  of homes, 41–61
Media Lab, 25

N

National Association of Housebuilders,
  1, 21
Naturally accountable, 193
Network economics, 68
Niche markets, definition of, 28

O

Office Information Systems, 6
Ontological security, 42
OSGI, Open standards Gateway Initiative,
  254
Optimality, 64
Orange, mobile operator, 1, 2, 3, 4
Orange at Home project, 1, 2, 4, 11,
  227–245

P

Paid and unpaid work, 64–66
Paperless offices, 102
Paper newspapers, 207, 211, 214, 215–6
Paper mail, 9, 101–12, 214
PDAs, 1, 3–4, 108, 111
Perceptual modalities, 122
Periodicities, 67
Personal
  distance, 132
  systems, definition of, 23
  Technologies, Journal of, 6
Personalisation, 235, 242–3, 259

R

Regionalisation, concept of, 41–42
  Goffman's concept, 93–4, 132
Residential gateway, 254

S

Semiology, 8
Smart Home, 17–29
  definitions of, 17, 21–22, 34–5,
    249
  concept of, 22, 31, 229–30
  house, 1
Social connectivity, 4, 227–245
Sociofugal spaces, 132
Sociopetal spaces, 132, 148,
Stanford Research Institute (SRI), 2
Substitution argument, 101–2
Systems approach, definition of, 43

T

Teletext, 81, 121
Time consuming goods, 66–7
  free time, 129–30
  localisation of, 129
  saving goods, definition of, 20–21
  using goods, definition of, 20–21,
    66–7
  shifting, 73
Technology push, 25, 29, 91, 163
  concept of, 22,
  determinism, 59–60
  imperative, 59
Television lounge, concept of, 20
Terminal manufacturers, 4
Time, economic category, 8

U

Ubiqitous computing, vision of, 183–184,
  199, 202, 204
Usage overhead, 232

V

VDI, (voice, data, image), definition of,
  250, 252
Viewing by appointment, 118
Video conferencing, 240, 244
Video connectivity, 244
  walls, 25

W

White goods, 7, 33
Workflow, 106–8, 11
Wirefree, 2